OSCILLATIONS IN NONLINEAR SYSTEMS

JACK K. HALE

DOVER PUBLICATIONS, INC.
MINEOLA, NEW YORK

Bibliographical Note

This Dover edition, first published by Dover Publications, Inc., in 2015, is an unabridged republication of the work first published by McGraw-Hill, New York, in 1963, and reprinted by Dover in 1992.

International Standard Book Number

ISBN-13: 978-0-486-67362-2
ISBN-10: 0-486-67362-6

Manufactured in the United States by Courier Corporation
67362602 2015
www.doverpublications.com

To Hazel and my parents

PREFACE

It is appropriate indeed that a monograph on the theory of oscillations be introduced with a tribute to the genius of Henri Poincaré and the enduring inspiration evoked by his treatises on celestial mechanics. Undoubtedly no mathematician of the nineteenth century can approach Poincaré's preponderance of ideas related to oscillatory phenomena, which have borne fruit in the works of others and are continuing to do so even to this day. For example, most of the important current research on the oscillatory behavior of conservative differential equations and behavior associated with the n-body problem is a direct outgrowth of results documented by Poincaré nearly a century ago.

Near the end of the last century, it was Lord Rayleigh who made what was a startling observation for his day and thereby contributed significantly to the theory of oscillations. In connection with his work on the theory of sound, he discovered a nonconservative second-order nonlinear differential equation which could sustain an oscillation without the influence of external disturbances. However, the significance of this discovery of self-excited oscillation was not realized immediately. It remained for van der Pol, in his work on the triode oscillator in the 1920s, to encounter and develop the theory for essentially the same equation discovered by Lord Rayleigh. Successful proof that all its nontrivial solutions approached a specified periodic solution was a significant step forward. Van der Pol also observed that the same basic equation could be used to describe a wide variety of interesting physical phenomena, and, in particular, demonstrated its use in describing certain types of irregularities in the beating of the heart.

The observations and results of van der Pol led to a renewed interest in problems concerning oscillatory phenomena in nonlinear differential equations. Motivated by an interest in understanding the behavior of the solutions of the now-labeled "van der Pol equation" subjected to external forces, Krylov and Bogoliubov in the early 1930s introduced their famous method of averaging. Further research in the last twenty-five years has led to the discovery of many more new and interesting types of oscillatory behavior.

It is the purpose of the present monograph to discuss some aspects of oscillatory behavior in nonlinear differential equations which depend very heavily upon the fact that the equations are nonlinear. Most of the discussion deals with equations which contain a small parameter. The reason for this is twofold. First of all, experience has indicated that a detailed discussion of equations with a small parameter reveals in many situations what is to be expected for arbitrary nonlinear equations. Secondly, with the exception of equations of order two, only results of a very specialized character are possible for arbitrary nonlinear equations. However, as will be demonstrated, a tractable theory with small perturbations is available.

In the hope that new applications may be discovered, a unified presentation of the results has been given in such a way that the underlying methods are easily understood. To avoid excessive mathematical sophistication and to make the material suitable for as large an audience as possible, the results are presented with stronger hypotheses than necessary.

Some familiarity with matrices, linear differential equations, and the concept of stability is required. In order to make the exposition reasonably self-contained, these basic concepts are discussed briefly in Part I. Except for Chapter 5, all of Part II depends upon the theory of Chapter 6. A good understanding of the examples in Chapter 7 will yield most of the ideas of Part II. Chapter 11 contains an abstraction of the technique used in Chapter 6 together with applications to other types of problems. An understanding of the properties of almost periodic functions given in Chapter 12 is necessary for all of Part III. Chapter 15 contains the theory of integral manifolds and averaging, and the remaining chapters are merely applications of these results.

The author wishes to acknowledge all comments and suggestions that have been made by all members of the staff at RIAS. A special thanks goes to Professor Nelson Onuchic, who carefully read most of the original manuscript. The comments by Professors Lamberto Cesari, Warren Loud, and Walter T. Kyner are greatly appreciated. Finally, part of the research which stimulated this monograph was supported by the Air Force Office of Scientific Research of the Office of Aerospace Research and the U.S. Army, Army Ballistic Missile Agency.

Jack K. Hale

CONTENTS

Part I

Introduction and Background Material

1: Introduction

Most physical systems are nonlinear. We shall assume the evolution of the physical system is governed by a real ordinary differential equation; that is, the state $x(t) = (x_1(t), x_2(t), \ldots, x_n(t))$ of the physical system at time t is a point along the solution of the differential system

$$\dot{x}_i = f_i(x_1, x_2, \ldots, x_n) \qquad i = 1, 2, \ldots, n \tag{1-1}$$

which passes through the point x_i^0; $i = 1, 2, \ldots, n$, at time $t = t_0$.

In general, the functions f_i are nonlinear functions of the state variables $x_1, x_2, \ldots, x_n$. For the sake of simplicity in analyzing (1-1), the functions f_i are frequently replaced by linear functions. In many cases this is sufficient, but there are phenomena which cannot be explained by analysis of the linear approximation.

The purpose of the present book is to concentrate on some aspects of differential equations which depend very strongly upon the fact that (1-1) is nonlinear.

The basic quality of a linear system (1-1) is (1) the sum of any two solutions of (1-1) is also a solution (the principle of superposition) and (2) any constant multiple of a solution of (1-1) is also a solution. Consequently, knowing the behavior of the solutions of (1-1) in a small neighborhood of the origin, $x_1 = x_2 = \cdots = x_n = 0$, implies one knows the behavior of the solutions everywhere in the state space; that is, globally. Furthermore, if one has a periodic solution of a linear system (1-1), then it cannot be isolated since any constant multiple of a solution is also a solution.

In nonlinear systems none of the above properties need be true. In fact, there is no principle of superposition, the behavior of solutions is generally only a local property, and there may be isolated periodic solutions (except for a phase shift). A simple example illustrating the local property of the behavior of solutions is

$$\dot{x} = -x(1 - x)$$

whose solutions are shown in Fig. 1-1.

The most classical example of a system which has an isolated periodic solution (except for a shift in phase) is the van der Pol equation

$$\ddot{x} - \epsilon(1 - x^2)\dot{x} + x = 0 \qquad \epsilon > 0 \tag{1-2}$$

whose trajectories in the $(x,\dot{x})$ plane are shown in Fig. 1-2. The

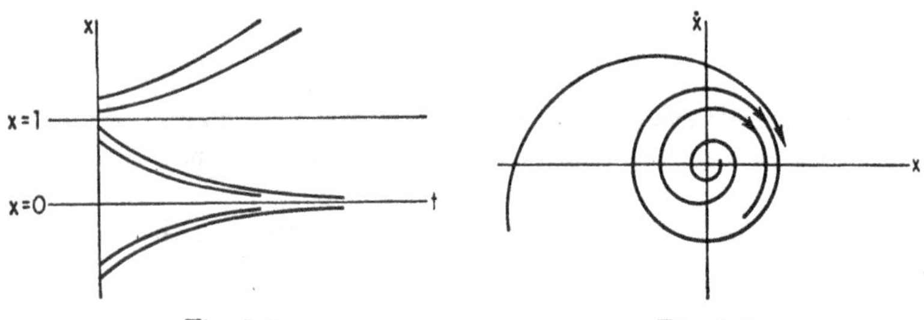

Fig. 1-1 **Fig. 1-2**

closed curve C has the property that all other trajectories approach it as $t \to \infty$ except, of course, the trajectory which passes through the equilibrium point $x = \dot{x} = 0$. This is a phenomenon which is due to the nonlinear structure of the system and could never be explained by a linear analysis. Such an oscillation is called *self-excited.*

Another interesting phenomenon that may occur in nonlinear systems is the following: Suppose system (1-1) is linear and apply a periodic forcing function of period T to (1-1). If the unforced system has no periodic solution, then there can never be an isolated periodic solution of any period except T. In nonlinear systems, this is not the case and isolated periodic solutions of period mT, where m is an integer greater than 1, may even occur. This phenomenon is known as *subharmonic resonance.*

Consider the forced van der Pol equation

$$\ddot{x} - \epsilon(1 - x^2)\dot{x} + x = \nu \cos \omega t \tag{1-3}$$

Is it possible to obtain solutions of this equation which oscillate with two basic frequencies, one due to the forcing function $\cos \omega t$ and one due to the basic frequency of the self-excited oscillation discussed before?

If such a phenomenon occurs, we say the solution has *combination tones*. As we shall see, such solutions may exist.

Our knowledge of nonlinear systems is still far from being complete. For the case where the system of differential equations has order 2 (that is, one degree of freedom), much more is known than for higher-order systems. The reason for this is that analytical-topological methods may be applied very nicely for systems of order 2, whereas for higher dimensions, the techniques of topology are not sufficiently

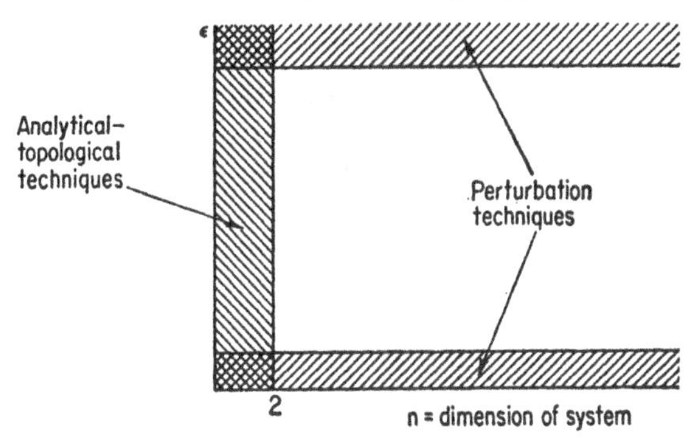

Fig. 1-3

developed. For systems of order greater than 2, the differential equations are usually assumed to contain a given parameter, and some type of perturbation technique is used to discuss the behavior of solutions. By using these perturbation techniques, one can build up a catalogue of phenomena which may occur in higher-order systems in the hope that one can use this experience as a guide to the eventual development of topological techniques which are applicable for higher-order systems. If we let the given parameter in our differential system be ϵ, then our knowledge of oscillatory phenomena for systems of differential equations lies almost entirely in the shaded regions of Fig. 1-3. As we shall see below, some techniques are available for discussing periodic solutions in the unshaded region of Fig. 1-3.

The techniques explained in this book will be methods of successive approximations and most of the emphasis will be on equations which contain a small parameter. The main reason for doing this is that the analytical-topological techniques would require more space than a book of this size, and also these techniques are covered very well in other places (see, for example, Cesari [1]* and Lefschetz [1]).

* Numbers in brackets indicate references listed in the Bibliography at the back of the book.

Chapters 2 to 4 are summaries of required material on matrices, linear systems of differential equations, and the basic stability theorems of Liapunov.

Part II is concerned mainly with periodic solutions of differential equations, although Chap. 11 does contain some material on generalized characteristic exponents. We attempt to give a unified presentation of the problems of finding periodic solutions of nonautonomous differential systems, autonomous differential systems, and characteristic exponents of linear periodic differential systems.

Many methods have been devised for solving these problems, and it is perhaps worthwhile to mention the general ideas involved. The reader who is not already familiar with at least one of these methods should most probably read Chap. 6 before attempting to read the discussion in the next few paragraphs.

Let us consider a nonautonomous differential system

$$\dot{x} = Ax + \epsilon f(t,x) \tag{1-4}$$

where x is an n vector, A is a constant matrix, f is periodic in t of period T, and ϵ is a small parameter. We wish to obtain a periodic solution of (1-4) of period T which, for $\epsilon = 0$, is a solution of the linear system

$$\dot{x} = Ax \tag{1-5}$$

Let φ_j, $j = 1, 2, \ldots, k$, be a maximal linearly independent set of periodic solutions of (1-5) of period T and for any constant vector $a = (a_1, \ldots, a_k)$, let

$$x_0(t,a) = a_1\varphi_1(t) + a_2\varphi_2(t) + \cdots + a_k\varphi_k(t) \tag{1-6}$$

The basic problem is to determine some vector $a = (a_1, \ldots, a_k)$ so that there is a corresponding periodic solution $x(t,a,\epsilon)$ of (1-4) of period T with $x(t,a,0) = x_0(t,a)$. This is accomplished by determining functions $x(t,a,\epsilon)$, $P_j(a,\epsilon)$, $j = 1, 2, \ldots, k$, which are defined for all t, $-\infty < t < \infty$, a in some set U, and $0 \leqq |\epsilon| \leqq \epsilon_0$, which have the following properties: for any a in U, $x(t,a,0) = x_0(t,a)$, $x(t,a,\epsilon)$ is periodic in t of period T and if there exists an $a = a(\epsilon)$ so that

$$P_j(a_1(\epsilon), \ldots, a_k(\epsilon), \epsilon) = 0 \qquad j = 1, 2, \ldots, k \tag{1-7}$$

then the periodic function $x(t,a(\epsilon),\epsilon)$ becomes a solution of (1-4). Thus the problem of existence of a periodic solution of (1-4) is reduced to the problem of finding a solution of a system of transcendental equations. Equations (1-7) are usually referred to as the *determining equations* or *bifurcation equations.* Most of the methods that have been devised for obtaining periodic solutions of (1-4) differ only in the manner in which these equations are derived. A few of these ideas are discussed below.

If one wishes to solve (1-4) by successive approximations, taking $x^{(0)} = x_0(t,a)$, and

$$\dot{x}^{(r+1)} = Ax^{(r+1)} + \epsilon f(t,x^{(r)}(t)) \qquad r = 0, 1, 2, \ldots \tag{1-8}$$

with each iterate $x^{(r)}(t)$ periodic in t of period T, then each step of the approximation consists in finding a periodic solution of the nonhomogeneous linear system

$$\dot{x} = Ax + g(t) \tag{1-9}$$

where $g(t)$ is periodic of period T.

Unfortunately, system (1-9) has a periodic solution of period T if and only if

$$\int_0^T \psi(t)g(t)\,dt = 0 \tag{1-10}$$

for every periodic function of period T which is a solution of the system

$$\dot{\psi} = -\psi A \qquad \text{(the adjoint equation)} \tag{1-11}$$

Consequently, the method of successive approximations (1-8) will not give the desired result. However, one might suspect that it is possible to modify the algorithm (1-8) by subtracting from $f(t,x^{(r)}(t))$ some factor which results in equation (1-10) being satisfied. The successive approximations then take the form

$$\dot{x}^{(r+1)} = Ax^{(r+1)} + \epsilon f(t,x^{(r)}(t)) - \epsilon M_r(t,a,x^{(r)}(t),\epsilon) \qquad r = 0, 1, 2, \ldots \tag{1-12}$$

where $M_r(t,a,x^{(r)},\epsilon)$ is to be determined so that (1-10) is satisfied with g replaced by $f(t,x^{(r)}) - M_r$. If the process (1-12) converges for ϵ small to a function $x(t,a,\epsilon)$, then $x(t,a,\epsilon)$ satisfies

$$\dot{x} = Ax + \epsilon f(t,x) - \epsilon M(t,a,x,\epsilon) \tag{1-13}$$

If, in addition, there exists a constant a such that the determining equations $M(t,a,x(t,a,\epsilon),\epsilon) = 0$, then $x(t,a,\epsilon)$ is a solution of (1-4).

One method for choosing the M_r in (1-12) began with the paper of Cesari [4] concerning linear systems with periodic coefficients. This method was subsequently developed by Cesari, Hale, Gambill, Fuller, and Thompson and involves carrying out the above process with

$$M_r(t,a,x^{(r)},\epsilon) = D_r(a,\epsilon)x^{(r)} \tag{1-14}$$

where the $D_r(a,\epsilon)$ do not depend on t. By proving that $x^{(r)}$, $D_r(a,\epsilon)$ converge to $x(t,a,\epsilon)$, $D(a,\epsilon)$, respectively, for ϵ small, the determining equations become

$$D(a,\epsilon) = 0$$

References to the papers of the above authors dealing with this process may be found in the book of Cesari [1]. Even though many results were obtained using this method, it has certain theoretical shortcomings and cannot be generalized directly to arbitrary nonlinear systems. A more elegant method has been devised by Cesari [3] which does generalize to arbitrary nonlinear systems. We shall return to this below.

If $\varphi_1, \ldots, \varphi_k$ is a basis of periodic solutions of period T of (1-5), then one can define a method of approximations (1-12) with

$$M_r(t,a,x^{(r)},\epsilon) = C_1^r(a,\epsilon)\varphi_1(t) + \cdots + C_k^r(a,\epsilon)\varphi_k(t) \qquad (1\text{-}15)$$

The $C_j^r(a,\epsilon)$ depend only on a, ϵ and can be chosen in such a way that (1-10) is satisfied with g replaced by $f(t,x^{(r)}) - M_r$. The convergence of the $C_j^r(a,\epsilon)$ to $C_j(a,\epsilon)$ for ϵ small implies that the determining equations are

$$C_j(a,\epsilon) = 0 \qquad j = 1, 2, \ldots, k \qquad (1\text{-}16)$$

This method has been discovered independently by Friedrichs [1], Lewis [1,2], Malkin [1], and Sibuya [1].

The procedure just outlined can be interpreted in another way. For simplicity only, assume that all solutions of (1-5) are periodic; that is, there are n linearly independent periodic solutions $\varphi_1, \ldots, \varphi_n$ of (1-5) of period T. Then e^{At} is periodic of period T and the transformation

$$x = e^{At}y$$

applied to (1-4) yields

$$\dot{y} = \epsilon e^{-At}f(t,e^{At}y) \overset{\text{def}}{\equiv} \epsilon q(t,y) \qquad (1\text{-}17)$$

For $\epsilon = 0$, (1-17) is self-adjoint. If we wish to solve this by successive approximations taking $y^{(0)} = a$, a given constant n vector, and

$$\dot{y}^{(r+1)} = \epsilon q(t,y^{(r)}) \qquad r = 0, 1, 2, \ldots \qquad (1\text{-}18)$$

with each iterate $y^{(r)}$ periodic in t of period T, then we must have

$$\begin{aligned} 0 &= \int_0^T q(t,y^{(r)}(t))\,dt \qquad (1\text{-}19) \\ &= \int_0^T e^{-At}f(t,e^{At}y^{(r)}(t))\,dt \end{aligned}$$

for each r. For the particular case under consideration, it is easy to see that the vector $C^r = (C_1^r, \ldots, C_n^r)$ defined in the previous paragraph is equal to the right-hand side of (1-19) except for a multiplicative constant. Since (1-19) will not necessarily be satisfied for every r, the algorithm (1-18) is modified to

$$\dot{y}^{(r+1)} = \epsilon q(t,y^{(r)}) - \frac{\epsilon}{T}\int_0^T q(t,y^{(r)}(t))\,dt \qquad r = 0, 1, 2, \ldots \quad (1\text{-}20)$$

The fact that $y^{(r)}$ converges to $y(t,a,\epsilon)$ for ϵ small implies that the determining equations are

$$\int_0^T q(t,y(t,a,\epsilon))\,dt = 0 \quad (1\text{-}21)$$

This method is a very special case of a general procedure developed by Cesari [3] and is the one that will be employed in this monograph. As mentioned earlier, Cesari's method also may be applied to arbitrary nonlinear systems, and a discussion of this point is contained in Chap. 11. In Chap. 11, we also discuss generalizations of these basic ideas to a perturbation problem considered by Bogoliubov and Sadovnikov [1], and the generalized characteristic exponents considered by Golomb [3].

If $y(t)$ is a periodic solution of (1-17) then $y(t)$ has a Fourier series

$$y(t) = \Sigma a_k e^{ik\omega t} \qquad T = \frac{2\pi}{\omega}$$

One could therefore determine a periodic solution of (1-17) by requiring that the coefficients a_k be of such a nature that all the coefficients of the periodic function

$$h(t) \overset{\text{def}}{\equiv} \dot{y}(t) - \epsilon q(t,y(t)) \sim \Sigma b_k e^{ik\omega t} \quad (1\text{-}22)$$

be equal to zero. This yields an infinite set of equations for the coefficients a_k. To obtain the determining equations, one shows that, for any given a_0, one can determine the a_k as functions of a_0, ϵ for ϵ small in such a way that all the Fourier coefficients of $h(t)$ in (1-22) vanish except the one corresponding to the mean value of $h(t)$, which is given by b_0. The vector b_0 is then a function of only a_0, ϵ and the determining equations are $b_0 = 0$. This procedure has been applied by Bass [1,2], Golomb [1,2] and Wasow [1]. Bass [1,2] has also considered a generalization of this procedure which is applicable in certain cases to equations not containing a small parameter.

Part III deals with the role of integral manifolds in nonlinear oscillations. In particular, in this part, the combination tones mentioned above are treated and the effect of high-frequency forcing functions on oscillatory behavior is discussed. In systems of autonomous differential equations of order greater than 2, examples are given to show that there may be interesting oscillatory behavior which is much more complicated than periodic phenomena. The treatment in Part III is based very strongly upon the work of Krylov, Bogoliubov, and Mitropolski (see the Bibliography for references). The method of averaging is explained and examples are given as illustrations of the method. After reading Part III, it will be apparent that the theory of integral manifolds and its application is not nearly completed and there remain many interesting unsolved problems.

In Part III, we do not discuss the important research on integral manifolds which has been done by Diliberto [1] and his collaborators mainly because it would require a complete redevelopment of the theory. There is some overlap between the results, and the interested reader is referred to the original papers, most of which are contained in volumes II and IV of the "Contributions to the Theory of Nonlinear Oscillations," Princeton University Press, Annals of Mathematics Studies, and the papers of Diliberto mentioned above.

Throughout this monograph, many examples are given but the physical motivation is not included. The examples are given only to illustrate some of the different types of phenomena that can occur in nonlinear differential equations. If one understands well the methods that are being employed, the author feels that many other types of phenomena can be discussed by using the same procedures.

This book does not claim to cover all the aspects of nonlinear oscillations. In fact, because of lack of space, we have omitted almost all reference to the important subject of relaxation oscillations (singular perturbations). The subject of asymptotic expansions and differential-difference equations is not even mentioned, nor are those problems in oscillations which deal with whether or not a solution has zeros.

In this monograph, a modest attempt has been made to present only some aspects of the theory of nonlinear oscillations in such a way that the machinery will become a basic tool for the engineer and the remaining problems will be a stimulus for the mathematician to continue to develop the theory.

2: Matrices

We summarize some results on matrices which will be used in the following. For a discussion of matrices, see S. Perlis [1]. Any rectangular array of real or complex numbers is called a matrix. If the array has m rows and n columns it is called an $m \times n$ matrix. If $m = n$, the matrix is called *square*. A convenient notation for an $m \times n$ matrix is

$$\begin{bmatrix} a_{11} & a_{12} & \cdots & a_{1n} \\ a_{21} & a_{22} & \cdots & a_{2n} \\ \cdot & & & \\ \cdot & & & \\ \cdot & & & \\ a_{m1} & a_{m2} & \cdots & a_{mn} \end{bmatrix}$$

The quantities a_{ij}, $i = 1, 2, \ldots, m$, $j = 1, 2, \ldots, n$, are called the *elements* or *coordinates* of the matrix. For a fixed i and j, we sometimes say a_{ij} is the (i,j)th element of the matrix and write

$$A = (a_{ij}, i = 1, 2, \ldots, m; j = 1, 2, \ldots, n)$$

The set of numbers a_{ij}, $j = 1, 2, \ldots, n$, is called the ith row of the matrix and the set a_{ij}, $i = 1, 2, \ldots, m$, is called the jth column. Matrices which have more than one row or one column will be denoted by capital Latin letters $A, B, C, \ldots$. Matrices consisting of only one column, that is, $m \times 1$ matrices, are usually called m vectors and will be denoted by small Latin letters $w, x, y, z, \ldots$. If x is an m vector with elements $x_1, \ldots, x_m$, then we shall denote x by $x = \text{col}\,(x_1, \ldots, x_m)$, where the symbol "col" is to remind us that it is a column. If a matrix consists of only one element; that is, it is 1×1 matrix, then we say it is a scalar.

An $n \times n$ square matrix A is said to be diagonal if $a_{ij} = 0$ for $i \neq j$. The sum of the diagonal elements of a square matrix A is called the trace of A and is denoted by tr (A); that is, if

$$A = (a_{ij}, i, j = 1, 2, \ldots, n)$$

then

$$\operatorname{tr}(A) = \sum_{i=1}^{n} a_{ii} \tag{2-1}$$

If A is a diagonal matrix with all diagonal elements equal to 1, then A is called the identity matrix and is usually denoted by I.

If A, B are $m \times n$ matrices whose (i,j)th elements are a_{ij}, b_{ij}, respectively, then we say $A = B$ if $a_{ij} = b_{ij}$ for all i, j. The matrix with all elements equal to zero will be denoted by 0.

If A, B, C, D are $m \times n$ matrices whose (i,j)th elements are a_{ij}, b_{ij}, c_{ij}, d_{ij}, respectively, and α is a scalar then, by $C = A + B$, $D = \alpha A = A\alpha$, we mean $c_{ij} = a_{ij} + b_{ij}$, $d_{ij} = \alpha a_{ij}$ for all i, j.

If $A = (a_{ij}, i = 1, 2, \ldots, m; j = 1, 2, \ldots, p)$ is an $m \times p$ matrix and $B = (b_{ij}, i = 1, 2, \ldots, p; j = 1, 2, \ldots, n)$ is a $p \times n$ matrix, then $C = AB$ designates the $m \times n$ matrix whose (i,j)th element is $c_{ij} = \sum_{k=1}^{p} a_{ik}b_{kj}$. Equality, addition, and multiplication defined as above have all the usual properties, except matrix multiplication is not necessarily commutative; that is, AB does not necessarily equal BA.

Given an $m \times n$ matrix

$$A = (a_{ij}, i = 1, 2, \ldots, m; j = 1, 2, \ldots, n)$$

the complex conjugate of A, denoted by $\bar{A}$, is the $m \times n$ matrix whose (i,j)th element is $\bar{a}_{ij}$, where $\bar{a}_{ij}$ is the complex conjugate of a_{ij}. The transpose of A is denoted by A' and is the $n \times m$ matrix whose (i,j)th element is a_{ji}; that is

$$A = \begin{bmatrix} a_{11} & a_{12} & \cdots & a_{1n} \\ a_{21} & a_{22} & \cdots & a_{2n} \\ \cdot & & & \\ \cdot & & & \\ \cdot & & & \\ a_{m1} & a_{m2} & \cdots & a_{mn} \end{bmatrix} \qquad A' = \begin{bmatrix} a_{11} & a_{21} & \cdots & a_{m1} \\ a_{12} & a_{22} & \cdots & a_{m2} \\ \cdot & & & \\ \cdot & & & \\ \cdot & & & \\ a_{1n} & a_{2n} & \cdots & a_{mn} \end{bmatrix} \tag{2-2}$$

If A is a square matrix and $A = A'$, then A is *symmetric*, and if $A = \bar{A}'$, then A is *Hermitian*. It is clear that $\overline{AB} = \bar{A}\bar{B}$, $(AB)' = B'A'$.

For any square matrix A, det A shall denote the determinant of A. We have $\det A' = \det A$, and if B, C are square matrices, then $\det (BC) = (\det B)(\det C)$. If $\det A \neq 0$, then A is called *nonsingular*, and there exists a matrix B such that $BA = AB = I$. Such a matrix B is called the *inverse* of A and is denoted by A^{-1}. Furthermore, $\det (A^{-1}) = (\det A)^{-1}$, and $(AB)^{-1} = B^{-1}A^{-1}$ if A, B are nonsingular square matrices. Also $(A')^{-1} = (A^{-1})'$.

If A is an $n \times n$ square matrix, then the *characteristic polynomial* of A is defined to be $\det (A - \lambda I)$. The zeros of $\det (A - \lambda I)$ are called the *eigenvalues* of the matrix A. If λ_j is an eigenvalue of A, then any vector u_j, with at least one nonzero component, which satisfies the equation

$$(A - \lambda_j I)u_j = 0$$

is called an *eigenvector* associated with the eigenvalue λ_j. By expanding the characteristic polynomial, one obtains

$$\det (A - \lambda I) = (-1)^n\lambda^n + c_1\lambda^{n-1} + \cdots + c_n \tag{2-3}$$

where

$$c_n = \det A \tag{2-4}$$

Since the eigenvalues λ_j, $j = 1, 2, \ldots, r \leqq n$, of A must satisfy $\lambda_1 \cdots \lambda_r = c_n$, it follows that A is nonsingular if and only if all the eigenvalues of A are different from zero.

Let A be an $n \times n$ matrix and let λ_j be an eigenvalue of A. Then $\det (A - \lambda I)$ can be written as

$$\det (A - \lambda I) = (\lambda - \lambda_j)^{\mu_j} P_j(\lambda)$$

where $P_j(\lambda)$ is a polynomial in λ and $P_j(\lambda_j) \neq 0$. The *multiplicity* of λ_j is defined as μ_j and the *nullity* of λ_j, denoted by ν_j, is defined to be the number of linearly independent eigenvectors associated with λ_j. It is known that $\nu_j \leqq \mu_j$ and it is easy to see that $\nu_j < \mu_j$ in some cases. In fact, let A be the 2×2 matrix

$$A = \begin{bmatrix} 0 & 1 \\ 0 & 0 \end{bmatrix}$$

The eigenvalue zero of A has multiplicity 2 and all the corresponding eigenvectors are given by $u = \operatorname{col}(a,0)$, $a \neq 0$, which implies the nullity is 1.

If λ is an eigenvalue of a square matrix A, we say that λ has *simple elementary divisors* if the multiplicity of λ is equal to the nullity of λ.

If A is an $n \times n$ matrix and P is an $n \times n$ matrix with $\det P \neq 0$, then the matrix $B = P^{-1}AP$ is said to be obtained from A by a similarity transformation, or A is *similar* to B. If A is similar to B then $\det A = \det B$ and $\det (A - \lambda I) = \det (B - \lambda I)$.

If A is a square matrix of the form

$$\begin{bmatrix} B_1 & 0 & \cdots & 0 \\ 0 & B_2 & \cdots & 0 \\ \cdot & & & \\ \cdot & & & \\ \cdot & & & \\ 0 & 0 & \cdots & B_k \end{bmatrix}$$

where each B_j, $j = 1, 2, \ldots, k$ is a matrix of some dimension and 0 represents a zero matrix of the appropriate dimension, then we shall write $A = \operatorname{diag}(B_1, \ldots, B_k)$, and say A is *block diagonal.*

The following result is very basic in the theory of matrices:

Theorem 2-1. Given an $n \times n$ matrix A with eigenvalues λ_j of multiplicity μ_j and nullity ν_j, $j = 1, 2, \ldots, r$, there is a matrix P (whose elements are complex numbers) with $\det P \neq 0$ such that

$$P^{-1}AP = \operatorname{diag}(C_1^{(1)}, \ldots, C_{\nu_1}^{(1)}, \ldots, C_1^{(r)}, \ldots, C_{\nu_r}^{(r)})$$

$$C_j^{(k)} = \begin{bmatrix} \lambda_k & 1 & 0 & \cdots & & 0 \\ 0 & \lambda_k & 1 & \cdots & & 0 \\ \cdot & & & & & \\ \cdot & & & & & \\ \cdot & & & & & \\ 0 & 0 & 0 & \cdots & \lambda_k & 1 \\ 0 & 0 & 0 & \cdots & 0 & \lambda_k \end{bmatrix} \tag{2-5}$$

$$j = 1, 2, \ldots, \nu_k, \quad k = 1, 2, \ldots, r$$

where each $C_j^{(k)}$ is an $n_j^{(k)} \times n_j^{(k)}$ matrix, $\sum_{j=1}^{\nu_k} n_j^{(k)} = \mu_k$.

Such a representation of a matrix A is known as a *Jordan canonical form* and it is unique except for a permutation of rows and columns. The matrices $C_j^{(k)}$, $j = 1, 2, \ldots, \nu_k$, are called the *companion matrices* of the eigenvalue λ_k. All the companion matrices $C_j^{(k)}$ of the eigenvalue λ_k will be diagonal if and only if the eigenvalue λ_k has simple elementary divisors. *The Jordan canonical form of A will be diagonal if and only if each of the eigenvalues of A has simple elementary divisors.*

It is sometimes convenient to have a canonical form whose elements are also real. For our purposes the following result is sufficient:

Theorem 2-2. Given an $n \times n$ matrix A with eigenvalues λ_j of multiplicity μ_j, $j = 1, 2, \ldots, r$, there exists a matrix P, whose elements are real, such that

$$P^{-1}AP = \text{diag}\,(D_1, \ldots, D_s)$$

where D_j is a real matrix such that, for some k, its eigenvalues are λ_k, $\bar{\lambda}_k$. We shall say that such a representation is a *real canonical form.*

If x is an n vector, $x = \text{col}\,(x_1, \ldots, x_n)$ then we define the norm of x, $\|x\|$, to be a nonnegative scalar function of x which satisfies

$$\|x\| = 0 \qquad \text{if and only if } x = 0 \tag{a}$$
$$\|\alpha x\| = |\alpha| \cdot \|x\| \qquad \text{for any number } \alpha \tag{b}$$
$$\|x + y\| \leqq \|x\| + \|y\| \tag{c}$$

If A is any $n \times n$ matrix, then the norm $\|A\|$ of A is the smallest number k such that, for any n vector x, $\|Ax\| \leqq k\,\|x\|$. If $C = AB$, where A, B are $n \times n$ matrices, then $\|C\| \leqq \|A\| \cdot \|B\|$, $\|A + B\| \leqq \|A\| + \|B\|$, and $\|\alpha A\| = |\alpha| \cdot \|A\|$ for any number α.

If $A(t)$ is a matrix whose elements are differentiable functions of t, then we define $\dot{A}(t) = dA(t)/dt$ to be the matrix whose elements are the derivatives with respect to t of the elements of $A(t)$. Similarly, one defines $\int_{t_0}^{t_1} A(t)\,dt$. We also say $A(t)$ is continuous in t if each of its elements is a continuous function of t.

For any given $n \times n$ matrix A, the matrix $\exp A$ or e^A is defined to be that matrix given by the series

$$e^A = I + A + \frac{A^2}{2!} + \frac{A^3}{3!} + \cdots \tag{2-6}$$

a series which is convergent for all matrices A. If A is similar to B, then e^A is similar to e^B. It is easy to prove that $e^A \cdot e^B = e^{A+B}$ if $AB = BA$. For any real number t, e^{At} is defined in the obvious way, and by direct computation on the above series, one shows that

$$\frac{d}{dt}\,(e^{At}) = Ae^{At} = e^{At}A \tag{2-7}$$

for any constant matrix A.

For a given integer n, the linear space of all n vectors, with addition and norm defined as above, will be denoted by E^n.

3: Linear Systems of Differential Equations

In this chapter, we summarize some well-known results concerning linear systems of differential equations. For complete details, see Coddington and Levinson [1]. Consider the nth-order system of linear differential equations

$$\dot{x}_i = \sum_{j=1}^{n} a_{ij}(t)x_j \qquad i = 1, 2, \ldots, n$$

or, if $x = \text{col}\,(x_1, \ldots, x_n)$, $A(t) = (a_{ij}(t), i, j = 1, 2, \ldots, n)$, the matrix equation

$$\dot{x} = A(t)x \tag{3-1}$$

where $A(t)$ is continuous in t for all $t \geqq t_0$. For any given n vector x_0, it is known there is a unique solution of (3-1) which for $t = t_0$ has the value x_0. This solution is denoted by $x(t,t_0,x_0)$. By a fundamental system of solutions $X(t)$ of (3-1) we mean an $n \times n$ matrix such that the columns of $X(t)$ are linearly independent solutions of (3-1). If, in addition, $X(t_0) = I$, the identity, we say $X(t)$ is the principal matrix solution of (3-1). If $X(t)$ is a fundamental system of solutions of (3-1), then any solution $x(t,t_0,x_0)$ is a linear combination of the columns of $X(t)$. In fact, since $\det X(t_0) \neq 0$, let b be the unique solution of the equations $X(t_0)b = x_0$. Then the function $X(t)b$ is a solution of (3-1) and, by the uniqueness theorem, must coincide with $x(t,t_0,x_0)$.

From the formula for the derivative of a determinant, one can derive the following basic result: If $X(t)$ is a matrix solution of (3-1), then

$$\frac{d}{dt}[\det X(t)] = [\text{tr}\, A(t)][\det X(t)]$$

where tr A is defined in formula (2-1) and thus

$$\det X(t) = [\det X(t_0)] \exp \int_{t_0}^{t} [\operatorname{tr} A(u)]\, du \tag{3-2}$$

This relation shows that one can obtain a fundamental system of solutions of (3-1) by only specifying the initial matrix $X(t_0)$ so that $\det X(t_0) \neq 0$.

Suppose $X(t)$ is a fundamental system of solutions of (3-1). From (3-2), $\det X(t)$ is always different from zero and $X^{-1}(t)$ exists. Since $X(t)X^{-1}(t) = I$, the identity, it follows that

$$\begin{aligned} 0 = \frac{d}{dt}[X(t)X^{-1}(t)] &= \left[\frac{d}{dt}X(t)\right]X^{-1}(t) + X(t)\frac{d}{dt}X^{-1}(t) \\ &= A(t)X(t)X^{-1}(t) + X(t)\frac{d}{dt}X^{-1}(t) \\ &= A(t) + X(t)\frac{d}{dt}X^{-1}(t) \end{aligned}$$

or $X^{-1}(t)$ satisfies the equation

$$\frac{d}{dt}X^{-1}(t) = -X^{-1}(t)A(t) \tag{3-3}$$

If y is a $1 \times n$ matrix, that is, a row vector, then the equation

$$\dot{y} = -yA(t) \tag{3-4}$$

is called the *adjoint equation* of (3-1).

3-1. LINEAR EQUATIONS WITH CONSTANT COEFFICIENTS

Consider the special case of (3-1),

$$\dot{x} = Ax \tag{3-5}$$

where A is a constant matrix. The matrix

$$X(t) = \exp A(t - t_0) \tag{3-6}$$

is a principal matrix of solutions of (3-5). In fact, from formula (2-7), $X(t)$ is a matrix solution and $X(t_0) = e^0 = I$. Consequently, to understand the behavior of the solutions of (3-5), it is sufficient to know the behavior of the matrix $\exp A(t - t_0)$.

If P is a constant nonsingular matrix, the change of coordinates

$$x = Py \tag{3-7}$$

yields the equivalent system of differential equations

$$\dot{y} = P^{-1}APy \tag{3-8}$$

that is, a system in which the coefficient matrix is similar to A. Consequently, from Theorem 2-1, we can assume that P has been chosen in such a way that $P^{-1}AP$ has the form given in formula (2-5). Since $\det P \neq 0$, to understand the nature of the solutions of (3-5), it is sufficient to discuss the solutions of a system of the form

$$\dot{w} = Cw \tag{3-9}$$

where w is a p vector and C is a $p \times p$ matrix given by

$$C = \begin{bmatrix} \lambda & 1 & 0 & \cdots & 0 \\ 0 & \lambda & 1 & \cdots & 0 \\ \cdot & \cdot & \cdot & & \cdot \\ \cdot & \cdot & \cdot & & \cdot \\ \cdot & \cdot & \cdot & & \cdot \\ \cdot & \cdot & \cdot & & 1 \\ 0 & 0 & 0 & \cdots & \lambda \end{bmatrix} \tag{3-10}$$

where λ is a complex number. This will be sufficient since $P^{-1}AP$ is a block diagonal matrix with each block on the diagonal of the form (3-10).

By direct evaluation in the series (2-6), one observes that

$$e^{Ct} = \begin{bmatrix} 1 & t & \dfrac{t^2}{2!} & \cdots & \dfrac{t^{p-1}}{(p-1)!} \\ 0 & 1 & t & \cdots & \dfrac{t^{p-2}}{(p-2)!} \\ 0 & 0 & 1 & \cdots & \dfrac{t^{p-3}}{(p-3)!} \\ \cdot & & & & \\ \cdot & & & & \\ \cdot & & & & \\ 0 & 0 & 0 & \cdots & 1 \end{bmatrix} e^{\lambda t} \tag{3-11}$$

Since $\exp C(t - t_0)$ is a principal matrix solution of (3-9), all the solutions of (3-9) must be linear combinations of the columns of (3-11).

The behavior of the solutions of (3-5) is obtained by (3-7). In particular, if all the eigenvalues of A have negative real parts, then all the solutions of (3-5) approach zero as $t \to \infty$. If all the eigenvalues of A have nonpositive real parts and the eigenvalues with zero real parts have simple elementary divisors, then the solutions of (3-5) are bounded for all $t \geqq t_0$. This last result is true because the companion matrices of an eigenvalue with zero real part would all have only one element; that is, the corresponding equation (3-9) would be a scalar equation and no powers of t would enter in (3-11).

3-2. LINEAR EQUATIONS WITH PERIODIC COEFFICIENTS

Consider the system of linear periodic differential equations

$$\dot{x} = A(t)x \tag{3-12}$$

where x is an n vector, $A(t)$ is an $n \times n$ matrix which is continuous in t and periodic in t of some period T; that is, $A(t + T) = A(t)$.

The fundamental result for systems (3-12) concerns the Floquet representation of a fundamental system of solutions of (3-12) as the product of a periodic matrix of period T and a solution matrix for a system with constant coefficients. This is stated more precisely in the following theorem.

Theorem 3-1. If $X(t)$ is a fundamental system of solutions of (3-12), then so is $X(t + T)$. Corresponding to each such fundamental system $X(t)$, there exists a nonsingular periodic matrix $Q(t)$ of period T and a constant matrix B such that

$$X(t) = Q(t)e^{tB} \tag{3-13}$$

For a proof of this theorem, see Coddington and Levinson [1].

Let us discuss some of the implications of this result. First of all, we observe that this implies there is at least one solution $x(t)$ of (3-12) such that

$$x(t + T) = \rho x(t) \tag{3-14}$$

for all t, where $\rho \neq 0$ is a convenient constant (real or complex). In fact, if $x(t)$ is a solution of (3-12), then there exists a constant n vector x_0 such that

$$x(t) = X(t)x_0 = Q(t)e^{tB}x_0 \tag{3-15}$$

since $X(t)$ is a fundamental system of solutions. If $x(t)$ is to satisfy (3-14), then a simple calculation shows that ρ and x_0 must satisfy the relation

$$(e^{TB} - \rho I)x_0 = 0$$

where I is the identity matrix. Consequently, if ρ is an eigenvalue of e^{TB} and x_0 a corresponding eigenvector, then the solution $x(t)$ defined by (3-15) has the desired property.

The eigenvalues of the matrix e^{TB} are called the *characteristic multipliers* of system (3-12), and we have shown that to each characteristic multiplier ρ of (3-12) there is at least one solution which satisfies (3-14). To each characteristic multiplier ρ of (3-12) we can define a *characteristic exponent* τ by the relation $\rho = e^{\tau T}$. Of course, the numbers τ are not uniquely defined. In fact, if $T = 2\pi/\omega$, then τ is determined only up to an integral multiple of ωi. A characteristic multiplier ρ will be called *simple* if its multiplicity as a zero of det $[e^{TB} - \rho I]$ is 1.

A simple argument shows that the characteristic multipliers ρ do not depend upon the particular fundamental system of (3-12) that is chosen.

The matrix e^{tB} is a fundamental system of solutions of the linear system with constant coefficients

$$\dot{y} = By \tag{3-16}$$

and thus the behavior of the solutions of system (3-12) is clear from (3-13) and the discussion of the previous pages.

Suppose $X(t)$, $X(0) = I$, the identity, is a fundamental system of (3-12). From (3-2), it follows that

$$\det X(T) = [\det X(0)] \exp \int_0^T \operatorname{tr} A(t)\, dt = \exp \int_0^T \operatorname{tr} A(t)\, dt$$

If $\rho_1, \ldots, \rho_m$, $m \leqq n$ are the characteristic multipliers of (3-12) (that is, the eigenvalues of e^{TB}) and $\tau_1, \ldots, \tau_m$ are the corresponding characteristic exponents, then by (3-13), $X(T) = e^{TB}$, and this implies the following important relations:

$$\begin{aligned} \exp(\tau_1 + \cdots + \tau_m)T &= \rho_1 \cdots \rho_m = \exp \int_0^T \operatorname{tr} A(t)\, dt \\ \tau_1 + \cdots + \tau_m &\equiv \frac{1}{T} \int_0^T \operatorname{tr} A(t)\, dt \qquad (\text{mod } \omega i) \end{aligned} \tag{3-17}$$

A number a is called *congruent to b modulo d* and written $a \equiv b \pmod{d}$ if there exists an integer m such that $a - b = md$.

The representation (3-13) also implies the following result.

Theorem 3-2. All solutions of (3-12) approach zero as $t \to \infty$ if and only if $|\rho_j| < 1$ for each characteristic multiplier ρ_j of (3-12). All solutions of (3-12) are bounded for $0 \leqq t < +\infty$ if and only if each characteristic multiplier ρ_j of (3-12) has $|\rho_j| \leqq 1$ and if $|\rho_j| = 1$, then ρ_j must have simple elementary divisors (that is, ρ_j as an eigenvalue of e^{TB} must have simple elementary divisors). System (3-12) has a periodic solution if and only if there exists a characteristic multiplier of (3-12) which is equal to 1. If there is a characteristic multiplier of

(3-12) which is equal to -1, then (3-12) has a periodic solution of period $2T$.

This result can also be phrased in terms of the characteristic exponents τ_j in an obvious way. Thus we see that the boundedness and stability properties of (3-12) are known if one can determine the characteristic exponents (or multipliers) of (3-12). However, Theorem 3-1 is only an existence theorem and says nothing about how to calculate the exponents. The calculation of characteristic exponents is in general very difficult, and there has been success in this direction only for general systems of degree 2 (see Magnus [1]) and for higher-order systems which contain a small parameter. This latter case will be discussed in Chap. 8, but we need to know some general properties of how the characteristic exponents depend upon a parameter.

If $A = A(t,\epsilon)$ in (3-12) is a continuous function of t, ϵ for all t, $-\infty < t < \infty$, and ϵ in some neighborhood U of $\epsilon = 0$ (ϵ may be complex), then the matrix $B = B(\epsilon)$ in (3-13) is a continuous function of ϵ in U. Since the characteristic multipliers $\rho_j = \rho_j(\epsilon)$ are eigenvalues of the matrix $e^{TB(\epsilon)}$, they can be considered as continuous functions of ϵ. Furthermore, one can choose a representation of the characteristic exponent $\tau_j(\epsilon)$ defined by $\rho_j(\epsilon) = \exp T\tau_j(\epsilon)$ in such a way that $\tau_j(\epsilon)$ is a continuous function of ϵ. If, for $\epsilon = 0$, $A(t,0)$ is a constant matrix whose eigenvalues are λ_j, then one can choose $\tau_j(0) = \lambda_j$ and then define $\tau_j(\epsilon)$ as a continuous function of ϵ. If $T = \dfrac{2\pi}{\omega}$, $A(t,0)$ constant, and $\tau_j(0) = \lambda_j + m\omega i$ for any fixed integer m, then one can also define $\tau_j(\epsilon)$ as a continuous function of ϵ. If the elements of $A(t,\epsilon)$ are analytic functions of ϵ, then each characteristic exponent $\tau_j(\epsilon)$ can also be considered as an analytic function of some fractional power of ϵ.

To obtain the above continuity properties of the characteristic exponents and multipliers as a function of a parameter, it is not necessary to assume $A(t,\epsilon)$ is continuous in both t and ϵ. In fact, one need only assume $A(t,\epsilon)$ is continuous in ϵ in some region U of $\epsilon = 0$ and there exists a function $\eta(t)$, L integrable in $[0,T]$ such that $\|A(t,\epsilon)\| \leqq \eta(t)$, $0 \leqq t \leqq T$, ϵ in U. A proof of this fact may be supplied, for example, by using the same type of reasoning as in Cesari and Hale [1].

As a final remark, if $A(t)$ in (3-12) is a real $n \times n$ matrix and ρ is a characteristic multiplier of (3-12), then the complex conjugate $\bar{\rho}$ is also a characteristic multiplier. In fact, if $X(t)$ is a real fundamental system of (3-12) with $X(0) = I$, the characteristic equation $\det [X(T) - \rho I] = 0$ is a polynomial equation with real coefficients.

4: Stability of Solutions of Nonlinear Systems

Consider a system of nonlinear differential equations

$$\dot{x} = q(x,t) \tag{4-1}$$

where t is a scalar, $x = \text{col}\,(x_1, \ldots, x_n)$, $q = \text{col}\,(q_1, \ldots, q_n)$, $q(x,t)$ is continuous in x, t together with its first and second partial derivatives with respect to x for all x, t with $\|x\| \leqq R$, $t \geqq t_0$. Hypotheses this strong on q are not necessary but are introduced only for simplicity.

Definition 4-1. A solution $x^*(t) = x(t,t_0,x_0)$ of (4-1) is said to be *stable* (to the right) if for every $\epsilon > 0$ there is a $\delta > 0$ such that:

1. If x_1 is such that $\|x_1 - x_0\| < \delta$, then every solution $x(t,t_0,x_1)$ of (4-1) exists for all $t \geqq t_0$ and $\|x(t,t_0,x_1)\| \leqq R$, $t \geqq t_0$.
2. If x_1 is such that $\|x_1 - x_0\| < \delta$, then

$$\|x(t,t_0,x_1) - x(t,t_0,x_0)\| < \epsilon \qquad t \geqq t_0$$

If $x^*(t) = x(t,t_0,x_0)$ is stable and, in addition, $\|x_1 - x_0\| \leqq \delta$ implies $\|x(t,t_0,x_1) - x(t,t_0,x_0)\| \to 0$ as $t \to \infty$, then x^* is said to be *asymptotically stable.* If $x^*(t)$ is not stable, then it is called *unstable.*

For any solution $x^*(t)$ of (4-1) the *linear variational equations* associated with x^* are defined to be

$$\dot{y} = Q(t)y \qquad Q(t) = \left(\frac{\partial q_j\,(x^*(t),t)}{\partial x_k}\; j,\, k = 1, 2, \ldots, n\right) \tag{4-2}$$

If $q(x,t)$ is periodic in t of period T and $x^*(t)$ is also periodic with the same period, then $Q(t + T) = Q(t)$.

The following theorem of Liapunov is well known, and a proof may be found in, for example, Cesari [1] or Lefschetz [1] or Coddington and Levinson [1].

Theorem 4-1. If $x^*(t)$ is a periodic solution of period T of (4-1) and $q(x,t)$ is periodic with the same period, and if all the characteristic exponents of (4-2) have negative real parts, then $x^*(t)$ is an asymptotically stable solution of (4-1). If one of the characteristic exponents of (4-2) has a positive real part, then x^* is an unstable solution of (4-1).

If system (4-1) is autonomous, that is,

$$\dot{x} = q(x) \tag{4-3}$$

then $x^*(t)$ being a solution of (4-3) implies $x^*(t + \varphi)$ is also a solution for every constant φ. Furthermore, $\dot{x}^*$ is a solution of (4-2). Consequently, if $x^*(t)$ is a nonconstant periodic solution of (4-3) of period T, then the periodic system (4-2) has a nonzero periodic solution and the previous theorem does not apply since one characteristic exponent is zero. In fact, from the remarks just made, one does not expect the strong type of stability mentioned above since one can always make a shift in phase φ to obtain another periodic solution of (4-3). However, in the n-dimensional space of the n vectors x, a periodic solution of an autonomous differential system defines a closed curve and one might suspect that, in some cases, this closed curve would be stable. We introduce the following definition.

Definition 4-2. Let $x^*(t)$ be a nonconstant periodic solution of (4-3), C be the closed curve defined by $x^*(t)$ in R^n, and $d(x,C) = \inf \|x - y\|$ for y on C. The solution $x^*(t)$ is said to be *orbitally stable* (to the right) if, for every $\epsilon > 0$, there exists a $\delta > 0$ such that, for every solution $x(t)$ of (4-3) with $d(x(t_0), C) < \delta$, we have $d(x(t), C) < \epsilon$ for all $t \geqq t_0$. If x^* is not orbitally stable, we say x^* is *unstable*. If x^* is orbitally stable and, in addition $d(x(t_0), C) < \delta$ implies $d(x(t), C) \to 0$ as $t \to \infty$, then we say x^* is *asymptotically orbitally stable*. If x^* is asymptotically orbitally stable and for every solution $x(t)$ with $d(x(t_0), C) < \delta$, there exists a constant φ such that $\|x(t) - x^*(t + \varphi)\| \to 0$ as $t \to \infty$, then x^* is called *asymptotically orbitally stable with asymptotic phase*.

The following theorem is also due to Liapunov.

Theorem 4-2. If $x^*(t)$ is a periodic solution of (4-3) and $n - 1$ of the characteristic exponents of (4-2) have negative real parts, then the solution x^* is asymptotically orbitally stable with asymptotic phase.

Part II

Periodic Solutions

5: Noncritical Cases

The purpose of the present chapter is to state and prove some basic results for the existence of periodic solutions for perturbed periodic differential systems. More specifically, we consider a system

$$\dot{y} = B(t)y + q(t,y,\epsilon) \tag{5-1}$$

where ϵ is a parameter, y and q are n vectors; B is an $n \times n$ matrix; B, q, are periodic and continuous in t of period T; q is also continuous in y, ϵ and is Lipschitzian with respect to y for $-\infty < t < \infty$, $0 \leqq \|y\| \leqq R$, $0 \leqq \epsilon \leqq \epsilon_0$. The linear system

$$\dot{y} = B(t)y \tag{5-2}$$

is said to be *noncritical* with respect to T, if it has no periodic solution of period T except the trivial solution $y = 0$. Otherwise, system (5-2) is said to be *critical*.

The theory for the existence of periodic solutions of (5-1) when (5-2) is *noncritical* is quite elementary. Although the noncritical case could be considered as a special case of a more general theory developed in Chap. 6, it is appropriate to discuss the noncritical case separately as an introduction to the problems since some of the lemmas below will be needed in later sections.

Lemma 5-1. If system (5-2) is noncritical with respect to T and $f(t)$, f an n vector, is any periodic function of period T, then there exists a unique function $y^*(t)$ which is periodic in t of period T and satisfies the equation

$$\dot{y} = B(t)y + f(t) \tag{5-3}$$

Furthermore, if $Y(t)$, $Y(0) = I$, the $n \times n$ identity matrix, is a fundamental system of solutions of (5-2), then

$$\|y^*(t)\| \leq \frac{K}{T}\int_0^T \|f(u)\|\,du$$
$$\frac{K}{T} = \sup_{0\leq t\leq T}\ \sup_{t\leq\tau\leq t+T} \|\{Y(\tau)[Y^{-1}(T) - I]Y^{-1}(t)\}^{-1}\| \tag{5-4}$$

a constant which is independent of f and depends only upon T and $Y(t)$.

Proof. It is clear that (5-3) has at most one periodic solution of period T. We now exhibit a periodic solution. If $Y(t)$ is the principal matrix solution of (5-2), then, from formula (3-3),

$$\frac{d}{dt}Y^{-1}(t) = -Y^{-1}(t)B(t)$$

Consequently, if $y(t)$ is a solution of (5-3) with $y(0) = y_0$, then

$$\frac{d}{dt}[Y^{-1}(t)y(t)] = Y^{-1}(t)f(t)$$

or

$$y(t) = Y(t)y_0 + \int_0^t Y(t)Y^{-1}(\tau)f(\tau)\,d\tau$$

If $y(t)$ is to be a periodic solution of (5-3) of period T; that is, $y(t + T) = y(t)$ for all t, then a simple calculation shows that y_0 is given uniquely by

$$y_0 = (Y^{-1}(T) - I)^{-1}\int_0^T Y^{-1}(u)f(u)\,du$$

and, thus, the periodic solution y^* of (5-3) is given by

$$\begin{aligned} y^*(t) &= Y(t)[Y^{-1}(T) - I]^{-1}\int_t^{t+T} Y^{-1}(u)f(u)\,du \\ &= \int_t^{t+T}\{Y(u)[Y^{-1}(T) - I]Y^{-1}(t)\}^{-1}f(u)\,du \end{aligned} \tag{5-5}$$

The above estimate (5-4) clearly holds since $y^*(t)$ is periodic, and the lemma is proved.

Theorem 5-1. Suppose system (5-2) is noncritical with respect to T and there exists a function $\eta(\epsilon,\rho)$, continuous and nondecreasing in ϵ, ρ for $0 \leq \epsilon \leq \epsilon_0$, $0 \leq \rho \leq R$, $\eta(0,0) = 0$, such that

$$\|q(t,y_1,\epsilon) - q(t,y_2,\epsilon)\| \leq \eta(\epsilon,\rho)\|y_1 - y_2\| \qquad q(t,0,0) = 0 \tag{5-6}$$

for $-\infty < t < \infty$, $\|y_1\| \leq \rho$, $\|y_2\| \leq \rho$, $0 \leq \epsilon \leq \epsilon_0$. Under these conditions, there exist constants $\sigma > 0$, $\epsilon_1 > 0$ such that there is a solution $y^*(t,\epsilon)$, $0 \leq \epsilon \leq \epsilon_1$, of (5-1), which is periodic in t of period T, continuous in ϵ at $\epsilon = 0$, $y^*(t,0) = 0$, and this solution is unique in the region $\|y\| \leqq \sigma$.

Proof. Consider the class S of all n-vector functions $y(t)$, periodic in t of period T, with the uniform topology; that is, if $y \in S$, the norm, $\nu(y)$ is defined as $\nu(y) = \sup_{0 \leqq t \leqq T} \|y(t)\|$. Let $S_\sigma = \{y \in S | \nu(y) \leqq \sigma\}$. For $y \in S_\sigma$ define the function $w = \mathfrak{F}y$ by the relation

$$w(t) = \mathfrak{F}y(t) = \int_t^{t+T} \{Y(u)[Y^{-1}(T) + I]Y^{-1}(t)\}^{-1} q(u,y(u),\epsilon)\, du \quad (5\text{-}7)$$

where $Y(u)$, $Y(0) = I$, is a fundamental system of (5-2). It is easy to verify that $w \in S$ and, if there exists a $y^* \in S$ such that $y^* = \mathfrak{F}y^*$, then y^* satisfies (5-1). The converse is also true from Lemma 5-1.

We wish now to show that, for σ sufficiently small and ϵ_1 sufficiently small, the operator $\mathfrak{F}$ maps S_σ into S_σ for each ϵ, $0 \leqq \epsilon \leqq \epsilon_1$, and is a contraction on S_σ (see Appendix).

From (5-6), there exists a continuous nondecreasing function $\mu(\epsilon)$, $0 \leqq \epsilon \leqq \epsilon_0$, $\mu(0) = 0$, such that

$$\begin{aligned} \|q(t,y,\epsilon)\| &\leqq \|q(t,y,\epsilon) - q(t,0,\epsilon)\| + \|q(t,0,\epsilon)\| \\ &\leqq \eta(\epsilon,\sigma)\|y\| + \mu(\epsilon) \end{aligned} \quad (5\text{-}8)$$

for all t, $-\infty < t < \infty$, $\|y\| \leqq \sigma$, $0 \leqq \epsilon \leqq \epsilon_0$. Let K be the constant defined in (5-4), and choose σ, $\epsilon_1 \leqq \epsilon_0$ so small that

$$K[\eta(\epsilon_1,\sigma)\sigma + \mu(\epsilon_1)] \leqq \sigma \qquad K\eta(\epsilon_1,\sigma) < 1$$

Then, clearly $\mathfrak{F}S_\sigma \subset S_\sigma$ for $0 \leqq \epsilon \leqq \epsilon_1$ and is a contraction on S_σ. Thus there is a unique fixed point $y^*(t,\epsilon)$ of $\mathfrak{F}$ in S_σ. The continuity of $y^*(t,\epsilon)$ in ϵ is easy to prove. To complete the proof, we show that $y^*(t,0) = 0$. Since $\mathfrak{F}$ is a contraction, $y^*(t,\epsilon)$ may be obtained as the uniform limit of the sequence $\{y^{(k)}(t)\}$ of functions

$$\begin{aligned} y^{(0)} &= 0 \\ y^{(k+1)}(t) &= \mathfrak{F}y^{(k)}(t) \qquad k = 0, 1, 2, \ldots \end{aligned}$$

From these relations, it is clear that $y^*(t,0) = 0$ and the theorem is proved.

The above proof of Theorem 5-1 is more involved than some in the literature (see Coddington and Levinson [1]), but we chose to present it in this way since it gives a simple method of successive approximations for the solutions, and it makes it more clear how the more complicated results of the next chapter generalize this method.

When the linear part of system (5-1), $\dot{y} = B(t)y$ contains a parameter, which may be either large or small, then some results similar to those of Theorem 5-1 can be obtained fairly easily when B is a constant. These are stated below.

Theorem 5-2. If B is a real constant $n \times n$ matrix, $\det B \neq 0$, if $q(t,y,\epsilon)$ satisfies relation (5-6), and if $h(t)$ is any continuous periodic function of period T with $\int_0^T h(t)\,dt = 0$, then there exist an $\epsilon_1 > 0$, $\epsilon_1 \leqq \epsilon_0$, and a $\sigma > 0$ such that the system of equations

$$\dot{y} = \epsilon[By + q(t,y,\epsilon) + h(t)] \tag{5-9}$$

has a periodic solution $y^*(t,\epsilon)$, $0 \leqq |\epsilon| \leqq \epsilon_1$, of period T, and this solution is unique in $\|y\| \leqq \sigma$ if $0 < |\epsilon| \leqq \epsilon_1$. Furthermore, $y^*(t,0) = 0$.

Proof. As in the proof of Theorem 5-1, let S be the space of periodic continuous functions of period T with norm ν and let $S_\sigma = \{y \in S | \nu(y) \leqq \sigma\}$. For $y \in S_\sigma$ define the function $w = \mathfrak{F}y$ by the relation

$$w(t) = \mathfrak{F}y(t) = (e^{-\epsilon BT} - I)^{-1}\epsilon \int_t^{t+T} e^{\epsilon B(t-u)}[q(u,y(u),\epsilon) + h(u)]\,du \tag{5-10}$$

Since $\det B \neq 0$, the matrix $e^{-\epsilon BT} - I$ is nonsingular if ϵ is $\neq 0$ and $|\epsilon| \leqq \epsilon_1$, $\epsilon_1 > 0$ is sufficiently small.

Integrating by parts in (5-10), it follows that

$$\begin{aligned} w(t) = \mathfrak{F}y(t) = \epsilon(e^{-\epsilon BT} - I)^{-1}\Big\{\epsilon B \int_0^T e^{-\epsilon Bv}\left[\int_t^{t+v} h(\tau)\,d\tau\right] dv \\ + \int_0^T e^{-\epsilon Bv}[q(t+v,\, y(t+v),\, \epsilon)]\,dv\Big\} \end{aligned}$$

since $\int_0^T h(u)\,du = 0$. Since $\lim_{\epsilon \to 0} \epsilon(e^{-\epsilon BT} - I)^{-1} = -(BT)^{-1}$, it follows that we can define

$$K = \sup_{0<|\epsilon|\leqq\epsilon_1} \|\epsilon(e^{-\epsilon BT} - I)^{-1}\|$$

For any $y \in S_\sigma$ and $\mu(\epsilon)$ defined as in (5-8), it follows that

$$\begin{aligned} \|w(t)\| &\leqq K\{|\epsilon|K_1 + [\eta(\epsilon,\sigma)\sigma + \mu(\epsilon)]K_2\} \\ \|w_1(t) - w_2(t)\| = \|\mathfrak{F}y_1(t) - \mathfrak{F}y_2(t)\| &\leqq KK_3\eta(\epsilon,\sigma)\nu(y_1 - y_2) \end{aligned}$$

for $0 \leqq t \leqq T$, $0 \leqq |\epsilon| \leqq \epsilon_1$ and some constants K_1, K_2, K_3.

Consequently, if $\epsilon_2 \leqq \epsilon_1 \leqq \epsilon_0$ and σ are chosen so small that

$$\begin{aligned} K\{|\epsilon|K_1 + [\eta(\epsilon,\sigma)\sigma + \mu(\epsilon)]K_2\} &\leqq \sigma \\ KK_3\eta(\epsilon,\sigma) &< 1 \end{aligned}$$

for $0 \leqq |\epsilon| \leqq \epsilon_2$, then $\mathfrak{F}$ maps S_σ into S_σ and is a contraction on S_σ. Thus there is a unique $y^*(t,\epsilon)$, $y^* \in S_\sigma$ such that $y^* = \mathfrak{F}y^*$; that is, y^* is a periodic solution of (5-9). Furthermore, this is the only periodic solution of (5-9) of period T for $0 < |\epsilon| \leqq \epsilon_2$, $\|y\| \leqq \sigma$ since the system $\dot{y} = \epsilon By$ is noncritical and the periodic solution must be a fixed point of $\mathfrak{F}$ (Lemma 5-1).

Finally, $y^*(t,\epsilon)$ is the uniform limit of the sequence of functions

$$y^{(0)} = 0$$
$$y^{(k+1)} = \mathfrak{F}y^{(k)}$$

with $\mathfrak{F}$ defined as in (5-10), and it is clear that $y^*(t,\epsilon) \to 0$ as $\epsilon \to 0$, completing the proof of Theorem 5-2.

An equivalent statement of Theorem 5-2 is the following.

Corollary 5-1. If B is a real constant $n \times n$ matrix, $\det B \neq 0$, if $q(t,y,\epsilon)$ satisfies relation (5-6), and if $h(t)$ is any continuous periodic function of period T with $\int_0^T h(t)\,dt = 0$, then there exist an $\omega_1 > 0$ and a $\sigma > 0$ such that the system of equations

$$\frac{dy}{dt} = By + q(\omega t,y,\epsilon) + h(\omega t) \tag{5-11}$$

has a periodic solution $y^*(t,\omega)$, $\omega \geqq \omega_1$, of period T/ω, and this solution is unique in $\|y\| \leqq \sigma$ for every finite $\omega \geqq \omega_1$. Furthermore,

$$\lim_{\omega \to \infty} y^*(t,\omega) = 0$$

Proof. If $\omega t = \tau$, $\epsilon = \omega^{-1}$, then system (5-11) is the same as (5-9) of Theorem 5-2 with $\omega^{-1} = \epsilon$.

Corollary 5-1 shows the influence of high-frequency forcing functions on a differential system. Under the assumption that the eigenvalues of B had negative real parts, Franell, Langenhop, and Levinson [1] first proved Corollary 5-1.

Theorem 5-3. If B is a constant $n \times n$ matrix whose eigenvalues have nonzero real parts and if $q(t,y,\epsilon)$ satisfies relation (5-6), then there exist an $\epsilon_1 > 0$ and a $\sigma > 0$ such that the system

$$\epsilon\dot{y} = By + q(t,y,\epsilon) \tag{5-12}$$

has a periodic solution $y^*(t,\epsilon)$ of period T, continuous in t, ϵ, $0 \leqq \epsilon \leqq \epsilon_1$, $y^*(t,0) = 0$, and $y^*(t,\epsilon)$ is unique in $0 \leqq \|y\| \leqq \sigma$.

Proof. As in the proof of Theorem 5-1, let S be the space of all periodic continuous functions of period T with norm ν and let $S_\sigma = \{y \in S | \nu(y) \leqq \sigma\}$. For $y \in S_\sigma$, define the function $w = \mathfrak{F}y$ by the relation

$$\begin{aligned} w(t) &= \mathfrak{F}y(t) \\ &= \left[\exp\left(\frac{-BT}{\epsilon}\right) - I\right]^{-1} \frac{1}{\epsilon} \int_t^{t+T} \left\{\exp\left[\frac{B(t-u)}{\epsilon}\right]\right\} q(u, y(u), \epsilon)\, du \end{aligned} \tag{5-13}$$

Since the eigenvalues of B have nonzero real parts, the matrix $e^{-BT/\epsilon} - I$ is nonsingular for every $\epsilon > 0$.

Without loss in generality we can assume that $B = \text{diag}\,(B_+, B_-)$ where B_+ is a square matrix which contains all the eigenvalues of B with positive real parts and B_- all those eigenvalues of B with negative real parts. If the vectors q, w are partitioned in the same way as $q = \text{col}\,(q_+, q_-)$, $w = \text{col}\,(w_+, w_-)$, then

$$\begin{aligned} w_+(t) &= \left[\exp\left(\frac{-B_+T}{\epsilon}\right) - I\right]^{-1} \\ &\qquad \times \frac{1}{\epsilon} \int_0^T \left[\exp\left(\frac{-B_+v}{\epsilon}\right)\right] q_+(t+v, y(t+v), \epsilon)\, dv \\ w_-(t) &= \left[I - \exp\left(\frac{B_-T}{\epsilon}\right)\right]^{-1} \\ &\qquad \times \frac{1}{\epsilon} \int_0^T \left\{\exp\left[\frac{B_-(T-v)}{\epsilon}\right]\right\} q_-(t+v, y(t+v), \epsilon)\, dv \end{aligned}$$

From the definitions of B_+, B_-, it follows that there exist constants K_1 and $\alpha > 0$ such that

$$\begin{aligned} \left\|\exp\left(\frac{-B_+v}{\epsilon}\right)\right\| &\leqq K_1 \exp\left(\frac{-\alpha v}{\epsilon}\right) & v \geq 0 \\ \left\|\exp\left[\frac{B_-(T-v)}{\epsilon}\right]\right\| &\leqq K_1 \exp\left[\frac{-\alpha(T-v)}{\epsilon}\right] & v \leqq T \\ \left\|\left[\exp\left(\frac{-B_+T}{\epsilon}\right) - I\right]^{-1}\right\| &\leqq K_1 \\ \left\|\left[I - \exp\left(\frac{B_-T}{\epsilon}\right)\right]^{-1}\right\| &\leqq K_1 \end{aligned}$$

for $0 < \epsilon \leqq \epsilon_0$.

For any $y \in S_\sigma$, and $\mu(\epsilon)$ defined as in (5-8)

$$\|w(t)\| \leqq \frac{2K_1^2}{\alpha}[\eta(\epsilon,\sigma)\sigma + \mu(\epsilon)]$$

$$\|\mathfrak{F}y_1(t) - \mathfrak{F}y_2(t)\| \leqq \frac{K_1}{\alpha}\eta(\epsilon,\sigma)\nu(y_1 - y_2)$$

for all t, ϵ, $0 \leqq t \leqq T$, $0 \leqq \epsilon \leqq \epsilon_0$. If σ, ϵ_1 are chosen so that

$$\frac{K_1}{\alpha}[\eta(\epsilon_1,\sigma)\sigma + \mu(\epsilon_1)] \leqq \sigma$$

$$\frac{K_1}{\alpha}\eta(\epsilon_1,\sigma) < 1$$

then the remainder of the proof follows as in the proof of Theorem 5-2, since we know that (5-2) has the solution $y = 0$ for $\epsilon = 0$ and B is nonsingular.

By combining the ideas used in the proofs of Theorems 5-1, 5-2, and 5-3 one can easily prove the following result, which contains as special cases each of the above theorems.

Theorem 5-4. Consider the system of equations

$$\begin{aligned} \dot{x} &= \epsilon[Ax + X(t,x,y,z,\epsilon) + h(t)] \\ \dot{y} &= B(t)y + Y(t,x,y,z,\epsilon) \\ \epsilon\dot{z} &= Cz + Z(t,x,y,z,\epsilon) \end{aligned} \tag{5-14}$$

where x, y, z are vectors, all functions are periodic in t of period T, continuous in t, x, y, z, ϵ, Lipschitzian in x, y, z for $-\infty < t < \infty$, $0 \leqq \|x\|, \|y\|, \|z\| \leqq R$, $0 \leqq \epsilon \leqq \epsilon_0$, and the Lipschitz constants of X, Y, Z approach zero as $\|x\|, \|y\|, \|z\|, \epsilon \to 0$. Furthermore, X, Y, Z all vanish when x, y, z, ϵ are equal to zero. If, in addition,

$$\int_0^T h(t)\,dt = 0$$

$\det A \neq 0$, the eigenvalues of C have nonzero real parts and the system $\dot{y} = B(t)y$ has no periodic solution of period T except $y = 0$, then there exist an $\epsilon_1 > 0$, $\sigma > 0$ such that (5-14) has a periodic solution for $0 < \epsilon \leqq \epsilon_1$, which for $\epsilon = 0$ is zero, and this solution is unique in the region $0 \leqq \|x\| + \|y\| + \|z\| \leqq \sigma$.

Systems of equations of the form (5-14) have been discussed by many authors. For example, with the equation in x absent, see Flatto and Levinson [1], Volosov [1], Anasov [1], Zadiraka [1], and with the equation in y absent, see Volk [1].

6: Periodic Solutions of Equations in Standard Form— Critical Cases

Consider the system of equations

$$\dot{z} = Az + \epsilon Z(t,z,\epsilon) \tag{6-1}$$

where ϵ is a parameter; t is a real variable; z, Z are n vectors; A is an $n \times n$ constant matrix; Z is periodic in t of period T, is continuous in t, z, ϵ, has a continuous first derivative with respect to ϵ and continuous first and second partial derivatives with respect to z for $-\infty < t < \infty$, $0 \leqq \|z\| \leqq R$, $0 \leqq |\epsilon| \leqq \epsilon_0$, $R > 0$, $\epsilon_0 > 0$. We shall briefly refer to system (6-1) satisfying all the above smoothness conditions as *system* (6-1). Such restrictive smoothness conditions on Z are not necessary and are assumed only for simplicity.

System (6-1) will be said to be in *standard form* if the matrix A has the form

$$A = \operatorname{diag}(0_p, B) \tag{6-2}$$

where 0_p is the $p \times p$ zero matrix and B is a $q \times q$ matrix, $q = n - p$, with the property that no solution of the equation

$$\dot{y} = By$$

is periodic of period T except the trivial solution $y = 0$. The system $z' = Az$ is critical since there are nontrivial constant solutions. This makes the problems different from those of the previous chapter.

As we shall see later, the problem of the existence of periodic solutions of weakly nonlinear differential systems and the determination of the characteristic exponents of almost constant linear periodic differential systems can be reduced to the determination of periodic solutions of systems of the form (6-1) which are in standard form.

In this chapter, we always assume that system (6-1) is in standard form and we want to find necessary and sufficient conditions for the existence of periodic solutions of (6-1), (6-2) of period T which, for $\epsilon = 0$, are periodic solutions of $\dot{z} = Az$ of period T. The most general periodic solution of period T of this system is $z = \text{col}\,(a,0)$ where a is an arbitrary constant p vector. The natural thing to do is to iterate on (6-1), (6-2) using as an initial approximation $z = \text{col}\,(a,0)$. Unfortunately, the successive interations will not in general be periodic, even though the final limit function may be periodic. The presence of these nonperiodic terms (the secular terms) in the successive approximations makes it very difficult to discuss the qualitative behavior of the solution. On the other hand, one does not expect to obtain a periodic solution starting with an arbitrary initial vector $\text{col}\,(a,0)$. Consequently, one attempts to determine the vector a in such a way that no secular terms appear in the successive iterations. This is the essence of most of the methods for obtaining periodic solutions by successive approximations. As a final remark, one would also hope to postpone as long as possible the determination of the vector a; for otherwise, some of the qualitative properties of the solutions may be obscured in the shuffle. More precisely, we shall reduce the problem of determining periodic solutions of (6-1), (6-2) to the determination of a p vector a, which is a solution of a set of transcendental equations (determining equations) (bifurcation equations).

For ease in presentation and the proof of the results, we formulate the method in terms of operators on a Banach space, but afterward we return to the successive approximations.

Let S be the space of all continuous n-vector functions of the real variable t, periodic in t of period T, with the uniform topology; that is, if $f \in S$, the norm $\nu(f)$ is given by $\nu(f) = \sup_t \|f(t)\|$. If $f \in S$, $f = \text{col}\,(g,h)$ where g is a p vector and h is an $n - p$ vector, we define the operator $P_0\colon S \to S$ so that

$$P_0 f = \text{col}\left[T^{-1} \int_0^T g(t)\,dt, 0\right]$$

that is, P_0 projects a function $f \in S$ into a constant n vector all of whose components are zero except the first p which are the mean values of the first p components of the vector f.

For given positive constants b, d, $b < d < R$, and a given constant p vector a, $\|a\| \leqq b$, we define the set $S_0 \subset S$ by the relation

$$S_0 = \{f \in S | P_0 f = \text{col}\,(a,0),\ \nu(f) \leqq d\} \tag{6-3}$$

Basic to our understanding of the general system (6-1) is the behavior of the solutions of the nonhomogeneous linear system for forcing functions in S. These results are embodied in Lemma 6-1.

Lemma 6-1. If $f \in S$, then the nonhomogeneous equation

$$\dot{z} = Az + f(t) \tag{6-4}$$

with A as in (6-1) has a periodic solution of period T if and only if $P_0 f = 0$. Furthermore, there is a unique periodic solution $z^*(t)$ of period T with $P_0 z^* = 0$. If this unique solution is designated by $z^*(t) = \int e^{A(t-u)} f(u)\, du$, then there exists a constant K which is independent of f such that

$$\|z^*(t)\| = \left\| \int e^{A(t-u)} f(u)\, du \right\| \leq \frac{K}{T} \int_0^T \|f(u)\|\, du \leq K\nu(f) \tag{6-5}$$

Proof. If $f \in S$, and $f = \text{col}\,(g,h)$, $z = \text{col}\,(x,y)$ where g, x are p vectors and h, y are $n - p$ vectors, then system (6-4) is equivalent to the system

$$\dot{x} = g(t)$$
$$\dot{y} = By + h(t)$$

Since the equation $\dot{y} = By$ is noncritical, Lemma 5-1 implies there is a unique periodic solution of the second equation above for every periodic function $h(t)$ and from formula (5-5), this solution $y^*(t)$ is given by

$$y^* = (e^{-BT} - I)^{-1} \int_0^T e^{-Bu} h(t + u)\, du$$

and $\|y^*(t)\| \leq (K/T) \int_0^T \|h(u)\|\, du$ for all t, where K is given in formula (5-4).

To analyze the equation $\dot{x} = g(t)$, it is sufficient to consider each component of this vector equation separately,

$$\dot{x}_j = g_j(t)$$

It is clear that, for each $j = 1, 2, \ldots, p$, this equation has a periodic solution if and only if $\int_0^T g_j(t)\, dt = 0$. Furthermore, there is only one solution x_j^* with $\int_0^T x_j^*(t)\, dt = 0$. In fact, if there were two then their difference would be a constant, and since each has mean value zero, this constant must be zero. It remains only to show that x_j^* can be bounded as stated in the lemma. It is sufficient to prove this for the real part of x_j^* since the imaginary part can be treated in the same way. Therefore, we assume $g_j(t)$ is real and prove that there exists a constant ξ_j which depends on g_j, $0 \leqq \xi_j \leqq T$ such that

$$x_j^*(t) = \int_{\xi_j}^{t} g_j(u)\, du$$

For any ξ, $0 \leqq \xi \leqq T$, the unique periodic solution $x_j^*(t)$ of mean value zero of our equation $\dot{x}_j = g_j(t)$ is given by

$$x_j^* = \int_{\xi}^{t} g_j(u)\, du - \frac{1}{T}\int_0^T \left[\int_{\xi}^{t} g_j(u)\, du\right] dt$$

Since x_j^* has mean value zero there must be a ξ_1, $0 \leqq \xi_1 \leqq T$ such that $x_j^*(\xi_1) = 0$. Choose $\xi = \xi_1$ in the above formula and the result follows.

This proves the lemma.

A few remarks should be made concerning Lemma 6-1. First of all, we have designated the unique periodic solution z^* of (6-4) with the integral operator in (6-5). However, this integral is very complicated since, as shown in the proof of Lemma 6-1, some of the limits of integration depend upon the function f. On the other hand, the estimate (6-5) holds and the sign $\int$ has most of the usual properties of integrals, which is sufficient for our purposes. A second remark concerns the manner in which one computes the function $z^*(t)$ in practical applications. If $z = \text{col}\,(x,y)$, $f = \text{col}\,(g,h)$ where x, g are p vectors, then the computation of z^* proceeds as follows:

1. If $g(t)$ has a Fourier series

$$g \sim \sum_{k \neq 0} a_k e^{ik\omega t} \qquad T = \frac{2\pi}{\omega}$$

then

$$x^* = \sum_{k \neq 0} \frac{a_k}{ik\omega} e^{ik\omega t}$$

Clearly, this is the value of x^* since it has mean value zero.

2. The computation of y^* is given by

$$y^* = (e^{-BT} - I)^{-1} \int_0^T e^{-Bu} h(t + u)\, du$$

On the other hand, if one assumes that B is a diagonal matrix (which in practice is usually the case), $B = \text{diag}\ (\lambda_1, \ldots, \lambda_{n-p})$, and

$h = \text{col}\,(h_1, \ldots, h_{n-p})$ has a Fourier series

$$h_j \sim \sum_k b_{kj} e^{ik\omega t}$$

then $$y_j^* = \sum_k \frac{b_{kj}}{ik\omega - \lambda_j} e^{ik\omega t} \qquad j = 1, 2, \ldots, n-p$$

This function satisfies

$$\dot{y}_j^* = \lambda_j y_j^* + h_j(t)$$

which is our equation for B in diagonal form.

Finally, we see that in most cases the computation of z^* is very easy and is obtained by integration of exponential functions forgetting about the constants of integration.

We now use this lemma to prove Theorem 6-1.

Theorem 6-1. For given constants $d > 0$, $0 < b < R$, there is an $\epsilon_1 > 0$ with the following property: corresponding to each constant p vector a, $\|a\| \leqq b$, and to each ϵ, $|\epsilon| \leqq \epsilon_1$, there is a unique function

$$z^*(t) = z(t,a,\epsilon) \in S_0$$

which has a continuous first derivative with respect to t, and whose derivative satisfies the relation

$$\dot{z}^* = Az^* + \epsilon Z(t,z^*,\epsilon) - \epsilon P_0 Z(t,z^*,\epsilon) \tag{6-6}$$

where A, $Z(t,z,\epsilon)$ are the functions given in (6-1), (6-2). Furthermore, $z(t,a,0) = a^*$, $a^* = \text{col}\,(a,0)$, and $z(t,a,\epsilon)$ has continuous first derivatives with respect to a, ϵ. Also, if $Z(t,z,\epsilon)$ has continuous first derivatives with respect to some parameter, then the function $z(t,a,\epsilon)$ also has continuous first derivatives with respect to this parameter. Finally $z^*(t)$ may be obtained by a method of successive approximations given by formula (6-9) below.

Remark. In the following, we shall sometimes refer to the function $z(t,a,\epsilon)$ of Theorem 6-1 as the *periodic function associated with system* (6-1), (6-2).

Proof of Theorem 6-1. Suppose the operator P_0 is defined as in the lines preceding formula (6-3), a is a given constant p vector,

$$a^* = \text{col}\,(a,0)$$

is an n vector, S_0 is defined as in (6-3), and the operator $\int$ is defined as in Lemma 6-1.

For any $z \in S_0$, we define $w = \mathfrak{F}z$ by the relation

$$w = \mathfrak{F}z = a^* + \epsilon\int e^{A(t-u)}(I - P_0)Z(u, z(u), \epsilon)\, du \tag{6-7}$$

where $a^* = \text{col}\,(a,0)$ and I is the identity operator.

It is obvious that $P_0\mathfrak{F}z = a^*$. Also, from (6-5),

$$\begin{aligned}\|\mathfrak{F}z(t)\| &\leqq \|a\| + 2|\epsilon|\frac{K}{T}\int_0^T \|Z(u, z(u), \epsilon)\|\, du \\ &\leqq b + |\epsilon|L_1\end{aligned}$$

where L_1 is a constant independent of $z \in S_0$. Therefore, there exists an $\epsilon_1' > 0$, $\epsilon_1' \leqq \epsilon_0$, such that $\nu(\mathfrak{F}z) \leqq d$ for $0 \leqq |\epsilon| \leqq \epsilon_1'$. Consequently $\mathfrak{F}S_0 \subset S_0$. Furthermore, if $x \in S_0$, $y \in S_0$, then

$$\begin{aligned}\|\mathfrak{F}x(t) - \mathfrak{F}y(t)\| &\leqq 2\epsilon\frac{K}{T}L_2\int_0^T \|x(u) - y(u)\|\, du \\ &\leqq 2\epsilon KL_2\nu(x - y)\end{aligned} \tag{6-8}$$

where L_2 is the Lipschitz constant for Z in $-\infty < t < \infty$, $\|z\| \leqq R$, $0 \leqq |\epsilon| \leqq \epsilon_0$. Thus there exists an $\epsilon_1 \leqq \epsilon_1'$ such that $\mathfrak{F}$ is a contraction for $0 \leqq |\epsilon| \leqq \epsilon_1$ and there is a unique fixed point $z^*(t) = z(t,a,\epsilon)$ of $\mathfrak{F}$; that is, $z^* = \mathfrak{F}z^*$. Clearly, z^* satisfies (6-6) and, from the assumptions on the matrix A in (6-1), (6-2), any function $z(t,\epsilon) \in S_0$, $0 \leqq |\epsilon| \leqq \epsilon_1$ and satisfying (6-6) must be a fixed point of $\mathfrak{F}$ and therefore is $z(t,a,\epsilon)$.

It is easy to see that $z(t,a,\epsilon)$ is Lipschitzian in a, by our assumption on Z and the fact that $\mathfrak{F}$ is a contraction.

Furthermore, the function $z^*(t) = z(t,a,\epsilon)$ is the uniform limit of the sequence of functions $\{z^{(k)}(t)\}$ defined by

$$\begin{aligned}z^{(0)} &= \text{col}\,(a,0) = a^* \\ z^{(k+1)}(t) &= \mathfrak{F}z^{(k)}(t) = a^* + \epsilon\int e^{A(t-u)}(I - P_0)Z(u, z^{(k)}(u), \epsilon)\, du \\ &\qquad k = 0, 1, 2, \ldots\end{aligned} \tag{6-9}$$

From (6-8), it is easy to see that

$$\nu(z^* - z^{(k)}) \leqq \left(\frac{\theta^k}{1 - \theta}\right)\nu[z^{(1)} - z^{(0)}]$$

where $\theta = 2\epsilon KL_2 < 1$, $z^*(t) = z(t,a,\epsilon)$.

We now show by induction that each of the functions $z^{(k)}(t)$ in (6-9) has continuous first derivatives with respect to a. We have $\partial z^{(0)}/\partial a = (I, 0)$. Assume that $z^{(k)}$ has continuous first derivatives with respect to a. Then

$$\frac{\partial z^{(k+1)}}{\partial a} = I + \epsilon\int e^{A(t-u)}(I - P_0)\left[\frac{\partial Z\,(u, z^{(k)}(u), \epsilon)}{\partial z}\right]\left[\frac{\partial z^{(k)}}{\partial a}\right] du \tag{6-10}$$

and, consequently $\dfrac{\partial z^{(k+1)}}{\partial a}$ is continuous. Let $a = (a_1, \ldots, a_p)$. Then there exists an $\epsilon_1'' \leqq \epsilon_1$ such that the sequence of functions $\left\{\dfrac{\partial z^{(k)}}{\partial a_j}\right\}$, $j = 1, 2, \ldots, n$, which necessarily belong to S, are uniformly bounded by a constant M, in $-\infty < t < +\infty$, $\|a\| \leqq b$, $0 \leqq |\epsilon| \leqq \epsilon_1'' \leqq \epsilon_1$.

From (6-10), we have

$$\left\| \frac{\partial z^{(k+1)}(t)}{\partial a_j} - \frac{\partial z^{(k)}(t)}{\partial a_j} \right\| = \left\| \epsilon \int e^{A(t-u)}(I - P_0) \left\{\left[\frac{\partial Z\ (u, z^{(k)}(u), \epsilon)}{\partial z} \right.\right.\right.$$
$$\left. - \frac{\partial Z\ (u, z^*(u), \epsilon)}{\partial z} \right] \frac{\partial z^{(k)}}{\partial a_j} + \left[\frac{\partial Z\ (u, z^*(u), \epsilon)}{\partial z} \right] \left(\frac{\partial z^{(k)}}{\partial a_j} - \frac{\partial z^{(k-1)}}{\partial a_j} \right)$$
$$\left.\left. + \left[\frac{\partial Z\ (u, z^*(u), \epsilon)}{dz} - \frac{\partial Z\ (u, z^{(k-1)}(u), \epsilon)}{dz} \right] \frac{\partial z^{(k-1)}}{\partial a_j} \right\} du \right\|$$
$$\leqq 2\epsilon M L_3 K[\nu(z^{(k)} - z^*) + \nu(z^{(k-1)} - z^*)]$$
$$+ 2\epsilon L_4 K \nu \left(\frac{\partial z^{(k)}}{\partial a_j} - \frac{\partial z^{(k-1)}}{\partial a_j} \right) \qquad j = 1, 2, \ldots, n$$

where L_3 is the Lipschitz constant for $\partial Z/\partial z$ and L_4 is a bound for $\partial Z/\partial z_j$, $j = 1, 2, \ldots, n$. Now let $\theta_1 = \max\ (2\epsilon M L_3 K, 2\epsilon L_4 K)$ and choose ϵ so small that $\theta_1 < 1$. Then by an easy induction process one finds that

$$\nu \left(\frac{\partial z^{(k+1)}}{\partial a_j} - \frac{\partial z^{(k)}}{\partial a_j} \right) \leqq K_1 \delta^k \nu(z^{(1)} - z^{(0)})$$

where K_1 is a convenient constant and $\delta = \max\ (\theta_1, \theta)$. This implies that the sequence $\{\partial z^{(k+1)}/\partial a_j\}$ converges uniformly to a continuous function $u'(t)$. But, from the uniform convergence of all functions involved, it follows that $u'(t) = \partial z\ (t,a,\epsilon)/\partial a_j$, and this shows that $\partial z\ (t,a,\epsilon)/\partial a_j$, $j = 1, 2, \ldots, n$, exists and is continuous.

The remainder of the proof of the theorem follows by using similar arguments and will not be repeated here.

Theorem 6-2. Let $z(t,a,\epsilon)$ be the function given by Theorem 6-1 for all $\|a\| \leqq b < R$, $0 \leqq |\epsilon| \leqq \epsilon_1$. If there exist an $\epsilon_2 \leq \epsilon_1$ and a continuous function $a(\epsilon)$ such that

$$P_0 Z(t,z(t,a(\epsilon),\epsilon),\epsilon) = 0 \qquad \|a(\epsilon)\| \leqq b < R \text{ for } 0 \leqq |\epsilon| \leqq \epsilon_2 \quad (6\text{-}11)$$

then $z(t,a(\epsilon),\epsilon)$ is a periodic solution of (6-1), (6-2) for $0 \leqq |\epsilon| \leqq \epsilon_2$. Conversely, if (6-1), (6-2) has a periodic solution $\bar{z}(t,\epsilon)$, of period T, continuous in ϵ, $\|\bar{z}(t,\epsilon)\| < R$, $0 \leqq |\epsilon| \leqq \epsilon_2 \leqq \epsilon_1$, then $\bar{z}(t,\epsilon) = z(t,a(\epsilon),\epsilon)$, where $z(t,a,\epsilon)$ is given in Theorem 6-1 and $a(\epsilon)$ satisfies (6-11). Therefore, for ϵ_2 sufficiently small, the existence of a continuous $a(\epsilon)$ satisfying (6-11) is a necessary and sufficient condition for the existence of a periodic solution of (6-1), (6-2) of period T.

Proof. The first part of the theorem is obvious. Now suppose that (6-1), (6-2) has a periodic solution $\bar{z}(t,\epsilon)$ for $0 \leqq |\epsilon| \leqq \epsilon_2$. Define $a(\epsilon)$ by $P_0\bar{z}(t,\epsilon) = \text{col}\,(a(\epsilon),0)$. Then $a(\epsilon)$ is continuous and $\|a(\epsilon)\| < R$. Hence, for some $b < R$, $\|a(\epsilon)\| \leqq b < R$ for $0 \leqq |\epsilon| \leqq \epsilon_2$. Since $\bar{z}(t,\epsilon)$ is a solution of (6-1), (6-2), we have

$$(I - P_0)(\dot{\bar{z}} - A\bar{z} - \epsilon Z(t,\bar{z},\epsilon)) = 0$$
$$P_0(\dot{\bar{z}} - A\bar{z} - \epsilon Z(t,\bar{z},\epsilon)) = 0$$

But, from the definition of P_0 and the form of A, $P_0(\dot{\bar{z}} - A\bar{z}) = 0$. Therefore, these equations reduce to

$$\dot{\bar{z}} = A\bar{z} + \epsilon(I - P_0)Z(t,\bar{z},\epsilon) \tag{6-12}$$
$$P_0 Z(t,\bar{z},\epsilon) = 0 \tag{6-13}$$

for $0 \leqq |\epsilon| \leqq \epsilon_2$. But from Theorem 6-1, there is a unique function $z(t,\epsilon)$ with $P_0 z(t,\epsilon) = \text{col}\,(a(\epsilon),0)$ which satisfies (6-12) and this function is given by $z(t,a(\epsilon),\epsilon)$ of Theorem 6-1 for $0 \leqq |\epsilon| \leqq \epsilon_1$. If we further restrict ϵ_2 so that $\epsilon_2 \leqq \epsilon_1$, then $\bar{z}(t,\epsilon) = z(t,a(\epsilon),\epsilon)$, $0 \leqq |\epsilon| \leqq \epsilon_2$. But since $\bar{z}(t,\epsilon)$ satisfies (6-13), $a(\epsilon)$ must satisfy (6-11) and the theorem is proved.

Before proceeding further, let us reinterpret the above theorems in terms of Fourier series and successive approximations. We wish to find a periodic solution of (6-1), (6-2) which for $\epsilon = 0$ is equal to col $(a,0)$, where a is a constant p vector. Let $z = (x,y)$, where x is a p vector,

$$x = a + \sum_{k \neq 0} a_k e^{ik\omega t}$$
$$y = \sum_k b_k e^{ik\omega t}$$

be a periodic function of period $T = 2\pi/\omega$. Then $P_0 z = \text{col}\,(a,0)$. To obtain a periodic solution of system (6-1), (6-2), of the above form, the natural procedure is to substitute these functions in (6-1) and require that the coefficient of the term in $e^{ik\omega t}$ vanish for all k. Theorem 6-1 asserts that for any p vector a, and ϵ sufficiently small, the coefficients a_k, b_k can be determined as functions of a, ϵ [by the method of successive approximations (6-9)] in such a way that system

(6-1) is satisfied except for a constant p vector. Theorem 6-2 asserts that, if the p vector a can be chosen so that equations (6-11) are satisfied, then one has a periodic solution of (6-1).

Equations (6-11) will be referred to as the *determining equations* for the periodic solution. Another term which has been applied in the literature to these equations is bifurcation equations. If we let Z in (6-1) be partitioned as $Z = \text{col}\,(X,Y)$, where X is a p vector, Y is a $q = n - p$ vector, then the determining equations (6-11) may be written more explicitly as

$$\frac{1}{T}\int_0^T X(t,z(t,a,\epsilon),\epsilon)\,dt = 0 \tag{6-14}$$

As an immediate consequence of the above procedure, we have the following result (compare this result with Theorem 5-1).

Theorem 6-3. If the equation $\dot{z} = Az$ has no periodic solution of period T except the trivial solution $z = 0$, then for ϵ sufficiently small, there always exists a periodic solution of (6-1) which approaches zero as $\epsilon \to 0$. Furthermore, in a neighborhood of $z = 0$, this solution is unique.

Proof. This is obvious since, in (6-2), $B = A$ and there are no auxiliary equations (6-11), (6-14) to satisfy.

Theorem 6-4. In system (6-1), (6-2), let

$$Z = \text{col}\,(X,Y) \qquad z = \text{col}\,(x,y)$$

where X, x are p vectors and define

$$X_0(x,y,\epsilon) = \frac{1}{T}\int_0^T X(t,x,y,\epsilon)\,dt \tag{6-15}$$

If there exists a p vector a_0, $\|a_0\| < R$, such that

$$X_0(a_0,0,0) = 0 \qquad \det\left[\frac{\partial X_0\,(a_0,0,0)}{\partial a}\right] \neq 0 \tag{6-16}$$

then there exist an $\epsilon_1 > 0$ and a periodic solution $z(t,\epsilon)$, $0 \leqq |\epsilon| \leqq \epsilon_1$, of system (6-1), (6-2) of period T with $z(t,0) = \text{col}\,(a_0,0)$.

Proof. With the vector Z, z partitioned as above, the determining equations (6-11), (6-14) for $\epsilon = 0$ are given by

$$X_0(a,0,0) = 0$$

where X_0 is defined in (6-15). The result then follows from the implicit-function theorem.

Theorem 6-4 is the same result that would be obtained from the periodicity conditions of Poincaré (see, for example, Coddington and Levinson [1]).

As we shall see many times in what follows, it is possible to obtain conditions for the existence of periodic solutions of (6-1) even when conditions (6-16) are not satisfied. Furthermore, these theorems are obtained by discussing the qualitative behavior of equations (6-11). The next result in this chapter is directed toward such theorems.

Before proceeding to this discussion notice that Theorem 6-2 concerns the existence of $a(\epsilon)$, but in practical problems (see Chaps. 7 to 10) one may want also to determine some other parameters as a function of ϵ. Clearly the above reasoning applies to this case. Furthermore, it is necessary in some problems to consider the period T as a function of ϵ, say $T = T(\epsilon)$. It is clear that the above procedure applies if we suppose $T(\epsilon)$ is continuous in ϵ and bounded for $0 \leqq |\epsilon| \leqq \epsilon_1$.

Definition 6-1. Consider a system of differential equations $z' = f(t,z)$ where z, f are n vectors. We say that this system possesses *property E with respect to Q* if there exists a nonsingular matrix Q such that

$$Q^2 = I \qquad Qf(-t,Qz) = -f(t,z) \qquad QP_0 = P_0Q \tag{6-17}$$

where P_0 is the operator defined before Lemma 6-1.

Notice that Q in (6-17) commutes with P_0 if and only if

$$Q = \text{diag}\,(Q_1,Q_2)$$

where Q_1 is a $p \times p$ matrix. Property E expresses some symmetry conditions on the function $f(t,z)$ involving evenness and oddness. This property is easily recognizable in many systems since it only means that the differential equation is unchanged by replacing t by $-t$ and making the transformation of variables $z = Qw$.

Theorem 6-5. Suppose $Q = \text{diag}\,(Q_1,Q_2)$ where Q_1 is a $p \times p$ diagonal matrix (then necessarily each diagonal element of Q_1 is either $+1$ or -1) and suppose that system (6-1), (6-2) has property E with respect to this Q for all ϵ. If a, $\|a\| \leqq b < R$, is a p vector, $a^* = \text{col}\,(a,0)$ is an n vector, chosen in such a way that $Qa^* = a^*$, then the function $z(t,a,\epsilon)$ of Theorem 6-1 satisfies the relation

$$Qz(-t,a,\epsilon) = z(t,a,\epsilon)$$

and, consequently,

$$Z(-t,z(-t,a,\epsilon),\epsilon) = -QZ(t,z(t,a,\epsilon),\epsilon) \tag{6-18}$$

Proof. Consider the set $S_0^* \subset S_0$ consisting of those functions such that $Qz(-t) = z(t)$. Then $a^* = Qa^*$ belongs to S_0^*. Furthermore, if $z \in S_0^*$, then, since $QA = -AQ$ we have $Qe^{At} = e^{-At}Q$ and if $\mathfrak{F}$ is the operator defined in (6-7), then

$$\begin{aligned} Q\mathfrak{F}z(-t) &= Qa^* - \epsilon Qe^{-At}\textstyle\int e^{Au}(I - P_0)Z(-u, z(-u), \epsilon)\, du \\ &= a^* - \epsilon e^{At}\textstyle\int e^{-Au}(I - P_0)QZ(-u, z(-u), \epsilon)\, du \\ &= a^* - \epsilon e^{At}\textstyle\int e^{-Au}(I - P_0)QZ(-u, Qz(u), \epsilon)\, du \\ &= a^* + \epsilon e^{At}\textstyle\int e^{-Au}(I - P_0)Z(u, z(u), \epsilon)\, du \\ &= \mathfrak{F}z(t) \end{aligned}$$

and, therefore, $\mathfrak{F}z \in S_0^*$. The remainder of the proof is obvious.

Theorem 6-6. Suppose the conditions of Theorem 6-5 are satisfied. If the jth element of the diagonal of the matrix Q_1 is $+1$, then the jth equation in the determining equations (6-11) [or (6-14)] is satisfied for every p vector a for which $Qa^* = a^*$, $a^* = \text{col}\,(a,0)$.

Proof. The result follows immediately from relation (6-18) and (6-14) since the integrand of the jth component of the vector in (6-14) is an odd function of t.

If system (6-1), (6-2) has property E with respect to some matrix Q, then Theorems 6-5 and 6-6 assert that, for a proper starting vector a^* in the method of successive approximations (6-9), many of the equations in (6-11) [or (6-14)] are zero. This allows one to obtain existence theorems for periodic solutions even in cases when relations (6-16) are not satisfied. Property E can be used to prove the existence of families of periodic solutions and also assists in the computation of isolated periodic solutions. In the following, many applications of Theorems 6-5 and 6-6 will be given, but to grasp some of the implications, let us state a very simple corollary of Theorem 6-6.

Corollary 6-1. Let $Q = \text{diag}\,(I,Q_1)$ where I is the $p \times p$ identity matrix, $Q_1^2 = I$, and suppose that (6-1), (6-2) has property E with respect to this Q for all ϵ. If a is an arbitrary constant p vector, $\|a\| \leq b$, then there exist an $\epsilon_1 = \epsilon_1(b)$ and a function $z(t,a,\epsilon) \in S_0$, $z(t,a,0) = \text{col}\,(a,0)$, which satisfies (6-1); i.e., there exists a p-parameter family of periodic solutions of period T of (6-1).

In particular, Corollary 6-1 implies that, for every p vector a, $\|a\| \leqq b < R$, there exists an $\epsilon_1 = \epsilon_1(b)$ such that the equation

$$\dot{z} = \epsilon Z(t,z,\epsilon) \qquad Z(-t,z,\epsilon) = -Z(t,z,\epsilon) \tag{6-19}$$

where Z is periodic in t of period T and z is a p vector, has a solution $z(t,a,\epsilon)$ which is periodic in t of period T, $0 \leqq |\epsilon| \leqq \epsilon_1(b)$, and $z(t,a,0) = a$, $\det [\partial z(t,a,0)/\partial a] = 1$. The initial values at $t = t_0$ of $z(t,a,\epsilon)$ are given by $z_0 = z(t_0,a,\epsilon)$. If $\|z_0\| \leqq b_1 < b$, then there exists an $\epsilon_2 \leqq \epsilon_1(b)$ such that, for $0 \leqq |\epsilon| \leqq \epsilon_2$, there is a one-to-one correspondence between a and z_0 and all solutions of (6-19) are periodic.

Remark. The continuity conditions imposed on system (6-1) above are too restrictive. The easiest condition to eliminate is the continuity in t. Actually all that is needed is integrability in the sense of Lebesque. The essential element of this extension is to show that (6-5) is satisfied. We actually state the theorems of Chap. 8 in this more general setting. A theorem analogous to Theorem 6-1 can also be proved under only a Lipschitz condition in z (see Fuller [1] and Cesari [2]). Of course, the function $z(t,a,\epsilon)$ will not be so smooth as stated in Theorem 6-1. This less restrictive condition causes particular difficulties in the analysis to follow because one cannot use the most elementary form of the implicit function theorem to solve equations (6-11). We shall not mention this case again but refer the reader to the above-mentioned papers. Another extension of the above results to certain classes of almost periodic functions can be made. The essential thing is to preserve relation (6-5), and one can do this provided that the modules of the set of frequencies of $f(u)$ do not have the origin or any of the eigenvalues of B as an accumulation point. Extensions along this line by a different approach have been made by Golomb [1] (see also Chap. 11).

Another obvious extension of the above results is to the case where B in (6-2) is a function of t, periodic of period T, and where it is assumed that the equation $y' = B(t)y$ has no periodic solution of period T except $y = 0$. One merely changes the notation e^{At}, e^{Bt}, etc., by the principal matrix solutions of $y' = Ay$, $y' = By$, etc.

7: Practical Methods of Computing a Periodic Solution and Examples

From Theorems 6-1 and 6-2 we know that we can obtain a periodic solution of (6-1) if we find the function $z(t,a,\epsilon)$ by the method of successive approximations in (6-9) and then solve the equations (6-11). However, from the practical point of view, one can obtain an approximation to $z(t,a,\epsilon)$ to a given order in ϵ and then see if it is possible to find a simple root of the determining equations (6-11) for this approximate function. If this can be done, then from the continuity properties of the function $z(t,a,\epsilon)$, it follows that, for ϵ sufficiently small, the determining equations will have an exact solution close to the approximate solution so obtained.

In case the function Z in (6-1) is analytic in z, ϵ then the procedure is extremely simple. In fact, one defines the method of successive approximations as follows:

$$z^0 = \text{col}\,(a,0) = a^*$$
$$z^{(k+1)}(t) \equiv a^* + \epsilon \int e^{A(t-u)}(I - P_0)Z(u, z^{(k)}(u), \epsilon)\, du \qquad (\text{trunc } \epsilon^{k+2})$$
$$k = 0, 1, 2, \ldots \qquad (7\text{-}1)$$

where (trunc ϵ^{k+2}) signifies that all terms of order higher than ϵ^{k+1} are ignored. Then for any given value of k, $k = 0, 1, 2, \ldots$, one can check to see if it is possible to solve the approximate determining equations

$$P_0 Z[t, z^{(k)}(t), \epsilon] \equiv 0 \qquad (\text{trunc } \epsilon^{k+1}) \qquad (7\text{-}2)$$

for some p vector a_0 [remember that $z^{(k)}(t)$ depends upon the vector a]. If equations (7-2) have a solution whose corresponding Jacobian matrix has a nonzero determinant, then the determining equations (6-11) will have a solution for ϵ sufficiently small and we then know from Theorem 6-2 that there is a periodic solution of (7-1) close to col $(a_0,0)$ for ϵ sufficiently small.

The method of successive approximations (7-1), (7-2) is not the usual type of power-series expansion in powers of the small parameter ϵ. Many of the classical procedures assume the solution z and the vector a are power series in ϵ and then successively determine the coefficients of the expansion of a in such a way as to obtain a periodic solution. The solution z is thus obtained as a power series in ϵ whose coefficients do not depend upon ϵ (see Minorsky [1]).

In the above procedure, the function $z(t,a,\epsilon)$ may be determined to any degree of accuracy as a power series in ϵ with the coefficients depending upon the parameter a. In other words, the dependence of the function z upon a may be clearly exhibited. Afterward, the parameter a is determined as a function of ϵ so that the determining equations are satisfied. The function z may then be expanded as a new power series in ϵ, but it may be more desirable to have the more compact form where the dependence on a is preserved.

We now give some examples to illustrate some of the different types of problems that occur in the theory of nonlinear oscillations and also show how these problems are reduced to obtaining periodic solutions of equations in the "standard" form (6-1), (6-2). Also, some of the examples have been chosen to illustrate how the property E in Definition 6-1 is useful. All the examples are scalar second-order equations of the form

$$\ddot{w} + \sigma^2 w = \epsilon f(t,w,\dot{w})$$

but written as a system of two first-order equations by letting $w = x_1$, $\dot{w} = x_2$. These examples are analyzed only through the first approximation in ϵ. If higher-order approximations are necessary to obtain the information desired, then the method of successive approximations (7-1), (7-2) may be applied in a straightforward manner although, in doing so, the computations may become rather laborious. No examples of higher-order approximations are given since no essentially new ideas are introduced.

In later chapters, we show how to reduce arbitrary systems to "standard" form, but if the examples below are clearly understood, this should be almost self-evident.

7-1. NONAUTONOMOUS VAN DER POL EQUATION

Consider the equation

$$\begin{aligned} \dot{x}_1 &= x_2 \\ \dot{x}_2 &= -x_1 + \epsilon(1 - x_1^2)x_2 + \epsilon p \cos(\omega t + \alpha) \end{aligned} \tag{7-3}$$

where x_1, x_2 are scalars, $\epsilon > 0$, $p \neq 0$, ω, α are real numbers and $\omega^2 = 1 + \epsilon\beta$, $\beta \neq 0$. We wish to investigate whether or not this equation possesses a periodic solution of period $2\pi/\omega$ for ϵ sufficiently small.

To apply the results of the previous chapter, (7-3) must be transformed to "standard" form. The transformation of van der Pol (see Minorsky [1]) is ready-made for this problem. In fact, van der Pol suggested introducing the new coordinates z_1, z_2 by the relation

$$\begin{aligned} x_1 &= z_1 \sin \omega t + z_2 \cos \omega t \\ x_2 &= \omega(z_1 \cos \omega t - z_2 \sin \omega t) \end{aligned} \tag{7-4}$$

to obtain the new differential equations in z_1, z_2:

$$\begin{aligned} \dot{z}_1 &= \frac{\epsilon}{\omega} [\beta x_1 + (1 - x_1^2)x_2 + p \cos (\omega t + \alpha)] \cos \omega t \\ \dot{z}_2 &= -\frac{\epsilon}{\omega} [\beta x_1 + (1 - x_1^2)x_2 + p \cos (\omega t + \alpha)] \sin \omega t \end{aligned} \tag{7-5}$$

where x_1, x_2 are given in (7-4) and $\beta = (\omega^2 - 1)/\epsilon$. Equation (7-5) is now a special case of system (6-1) with $z = \text{col}\,(z_1,z_2)$ and the matrix A equal to the zero matrix. Therefore, we may apply our theory. In particular, since our equation (7-5) is analytic in ϵ, x_1, and x_2 for all x_1, x_2 and ϵ sufficiently small, we may apply (7-1) and (7-2). For $k = 0$, $z^{(0)} = a = \text{col}\,(a_1,a_2)$, and equations (7-2) for $k = 0$ are

$$\begin{aligned} \frac{\beta}{\omega^2} a_2 + \left(1 - \frac{a_1^2 + a_2^2}{4}\right) a_1 + \frac{p}{\omega^2} \cos \alpha = 0 \\ -\frac{\beta}{\omega^2} a_1 + \left(1 - \frac{a_1^2 + a_2^2}{4}\right) a_2 + \frac{p}{\omega^2} \sin \alpha = 0 \end{aligned} \tag{7-6}$$

If these equations have a solution a_{10}, a_{20} and the corresponding Jacobian of the left-hand sides of these equations has a nonzero determinant, then, for ϵ sufficiently small, there is an exact solution of the determining equations (6-11) and thus a periodic solution of (7-3) (see Theorem 6-4).

To analyze these equations, let $a_1 = r \cos \theta$, $a_2 = r \sin \theta$, to obtain the equivalent equations

$$\begin{aligned} r &= \frac{p}{\beta} \sin (\alpha - \theta) \\ \frac{\beta}{\omega^2} \cot (\alpha - \theta) + 1 &= \frac{p^2}{4\beta^2} \sin^2 (\alpha - \theta) \end{aligned} \tag{7-7}$$

Suppose that $0 < \beta \leqq p/2$ (equivalently $\epsilon^2 p^2/4\,(\omega^2 - 1) \geqq 1$) and let $\xi = \alpha - \theta, 0 \leqq \xi \leqq \pi/2$. Then there is a unique solution $\xi(\beta)$ of the last equation in

(7-7). If $y(\beta) = (1/\beta)\sin\xi(\beta)$, then the last equation in (7-7) is equivalent to

$$\frac{(1 - \beta^2 y(\beta))^{1/2}}{\omega^2 y(\beta)} + 1 = \frac{p^2}{4} y^2(\beta),\ 0 < y(\beta) \leqq \frac{1}{\beta} \tag{7-7A}$$

Using this relation, it is possible to show that $y(\beta)$ is monotone decreasing and $y(\beta) \to 2/p$ as $\beta \to p/2$ from below. Also, $y(\beta)$ approaches a limit $y^* = y^*(p, \omega)$ as $\beta \to 0$ from above. The amplitude $r = py(\beta)$ is the unique positive solution of the cubic equation $r^3 - 4r - (4p/\omega^2) = 0$.

The symmetry in β implies that the amplitude r as a function of β is the one shown in Fig. 7-1.

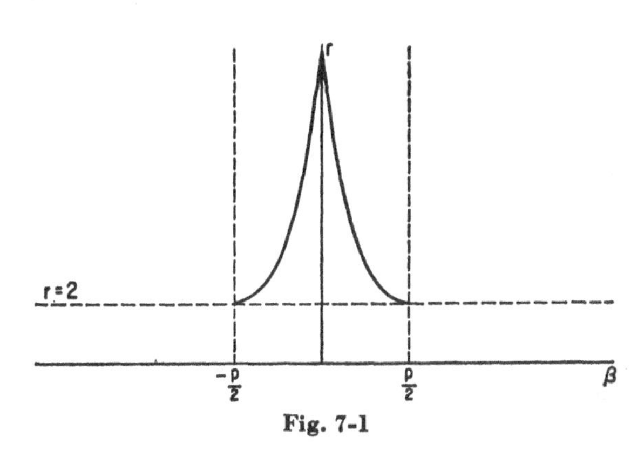

Fig. 7-1

For $\epsilon = 0$, system (7-3) has a free frequency which is equal to 1; that is, all the solutions are periodic of period 2π. On the other hand, if $\epsilon^2p^2/4(\omega^2 - 1)^2 \geqq 1$ for all ϵ, $0 \leqq \epsilon \leqq \epsilon_0$, sufficiently small, then we have seen that there is a periodic solution of (7-3) with frequency ω and an amplitude that is not small in ϵ even though the forcing function is small. In other words, if $\omega^2 - 1$ is of order ϵ, then the free frequency has been "locked" with the forcing frequency. This phenomenon is sometimes referred to as the *locking-in phenomenon* or *entrainment of frequency* (see Loud [3]). For more results on this equation, see Cartwright [1, p. 202] and Gambill and Hale [1, p. 385].

7-2. NONLINEAR MATHIEU EQUATION

Consider the nonlinear Mathieu equation

$$\begin{aligned} \dot{x}_1 &= x_2 \\ \dot{x}_2 &= -x_1 + \epsilon\alpha(\cos 2t)x_1 + \epsilon b x_1^3 \end{aligned} \tag{7-8}$$

where α, b are constants, and ϵ is a small parameter. In the previous example, we obtained a periodic solution whose period was the same as that of the forcing function, but we take as our problem here to try to find a periodic solution of (7-8) with a period of 2π; i.e., the period is twice that of the periodic terms in the equation itself. Such a phenomenon may be possible since the period of the periodic solutions of the unperturbed equation (i.e., $\epsilon = 0$) is in rational ratio with the period of the coefficients in (7-8).

As in the previous example, the transformation

$$\begin{aligned} x_1 &= z_1 \sin t + z_2 \cos t \\ x_2 &= z_1 \cos t - z_2 \sin t \end{aligned} \tag{7-9}$$

yields the equivalent system

$$\begin{aligned} \dot{z}_1 &= \epsilon[\alpha \cos 2t + b(z_1 \sin t + z_2 \cos t)^2](z_1 \sin t + z_2 \cos t) \cos t \\ \dot{z}_2 &= -\epsilon[\alpha \cos 2t + b(z_1 \sin t + z_2 \cos t)^2](z_1 \sin t + z_2 \cos t) \sin t \end{aligned} \tag{7-10}$$

which, of course, is a special case of system (6-1) with $z = \text{col}\,(z_1,z_2)$ and A equal to the zero matrix. Furthermore, the period of the coefficients in (7-10) is 2π, as desired.

We want to make use of Theorem 6-5 to simplify some of the calculations in this example. It is easily verified that system (7-10) has property E (see Definition 6-1) with respect to the matrix

$$Q = \text{diag}\,(-1,+1)$$

Let $a^* = \text{col}\,(0,a)$ where a is any given real number. Then $Qa^* = a^*$. Furthermore, let $z = \text{col}\,(z_1,z_2)$, $Z = \text{col}\,(Z_1,Z_2)$, where $Z_1(t,z)$ designates the function on the right-hand side of the first equation in (7-10) and Z_2 the function in the second equation. If $z(t,a,\epsilon)$ designates the function given by Theorem 6-1 associated with this scalar a, then it follows from Theorem 6-5 that $Z_2(t,z(t,a,\epsilon))$ is an odd function of t; that is, the second equation in the determining equations (6-11), or (6-14) is identically zero for every a. Consequently, it is only necessary to investigate the first equation.

Since our system (7-8) is analytic in ϵ, x_1, x_2, we can use (7-1) and (7-2). For $k = 0$, $z^{(0)} = \text{col}\,(0,a)$, we have that the equation in (7-2) reduces to the single equation

$$\frac{a}{8}\,(2\alpha + 3ba^2) = 0 \tag{7-11}$$

and from the remarks above, we need only investigate the simple zeros of (7-11).

It is clear that $a = 0$ corresponds to the zero solution of (7-8), which is of no interest. On the other hand, if $b\alpha < 0$, there exist two simple solutions of (7-11) given by $a = \pm(-2\alpha/3b)^{1/2}$ and these correspond to periodic solutions of (7-8) (see Theorem 6-4). The case $b\alpha > 0$ is reduced to the previous case by letting $t = \tau + \pi/2$.

For this simple example, it is not too much more difficult to forget about the symmetry properties of (7-10), but in more complicated examples, the above procedure is very efficient and in some cases even yields results that are not obtainable by the straightforward approach.

7-3. DUFFING'S EQUATION

Consider the Duffing equation

$$\begin{aligned} \dot{x}_1 &= x_2 \\ \dot{x}_2 &= -\sigma^2 x_1 - \epsilon\gamma x_1^3 + B\cos\omega t \end{aligned} \tag{7-12}$$

where σ, ϵ, γ, B, ω are parameters. If $p = k/m$, where k, m are positive integers and have no common divisor except 1, then one might suspect that, under certain conditions, if σ is close to $p\omega$, there might be a periodic solution of (7-12) whose dominant term is a sum of two terms, one of period $2\pi/\omega$ and the other of period $2\pi/p\omega$. If such a solution exists, we say it is a *subharmonic solution of order p*. We want to discuss the existence of such solutions.

For $\epsilon = 0$, system (7-12) has the particular solution $x_1 = A\cos\omega t$, $x_2 = -A\omega\sin\omega t$, $A = B/(\sigma^2 - \omega^2)$ if $\sigma \neq \omega$. Consequently if $\sigma \neq \omega$, the transformation

$$\begin{aligned} x_1 &= z_1\sin p\omega t + z_2\cos p\omega t + A\cos\omega t \\ x_2 &= p\omega(z_1\cos p\omega t - z_2\sin p\omega t) - A\omega\sin\omega t \end{aligned} \tag{7-13}$$
$$p = \frac{k}{m} \qquad A = \frac{B}{\sigma^2 - \omega^2}$$

applied to (7-12) yields the equivalent system

$$\begin{aligned} \dot{z}_1 &= -\frac{\epsilon}{p\omega}[\beta(z_1\sin p\omega t + z_2\cos p\omega t) \\ &\qquad + \gamma(z_1\sin p\omega t + z_2\cos p\omega t + A\cos\omega t)^3]\cos p\omega t \\ \dot{z}_2 &= \frac{\epsilon}{p\omega}[\beta(z_1\sin p\omega t + z_2\cos p\omega t) \\ &\qquad + \gamma(z_1\sin p\omega t + z_2\cos p\omega t + A\cos\omega t)^3]\sin p\omega t \end{aligned} \tag{7-14}$$

where $\beta = (\sigma^2 - p^2\omega^2)/\epsilon$.

Let us now apply the theory of Chap. 6 to this system (7-14). System (7-14) has property E with respect to $Q = \text{diag}\,(-1,1)$. Consequently, if we let $a^* = \text{col}\,(0,b)$, where b is an arbitrary constant, then $Qa^* = a^*$. From Theorem 6-6, it follows that the second determining equation in (6-14) is zero for every b and thus it only remains to discuss the first determining equation.

There are a few distinct cases to consider. If $p \neq 1, \frac{1}{3}$, and 3, then it follows immediately from (7-2) with $z^{(0)} = \text{col}\,(0,b)$ that the first approximation to the first determining equation is given by

$$\sigma^2 - p^2\omega^2 + \frac{3\epsilon\gamma}{4}(2A^2 + b^2) = 0$$

It is clear that, for any given b, one can always determine the forcing frequency ω in such a way as to solve the complete determining equations and thus obtain a subharmonic solution of order p.

The solution ω of the determining equations expressed as a function of b will be called the *frequency-response curve.*

Since $A = B/(\sigma^2 - \omega^2)$, it follows that the approximate frequency-response curve is

$$\omega^2 = \frac{\sigma^2}{p^2}\left\{1 + \frac{3\epsilon\gamma}{4\sigma^2}\left[\frac{2p^4B^2}{\sigma^4(1-p^2)^2} + b^2\right]\right\} \tag{7-15}$$

These curves are shown in Fig. 7-2 for $\epsilon\gamma > 0$ and $\epsilon\gamma < 0$. If $\epsilon\gamma > 0$, the minimum value of ω is attained for $b = 0$ and if $\epsilon\gamma < 0$, the maximum value of ω is attained for $b = 0$.

If $p = \frac{1}{3}$, then the approximate determining equation is

$$\sigma^2 - \frac{\omega^2}{9} + \frac{3\epsilon\gamma}{4}(b^2 + 2A^2 + bA) = 0$$

Again, it is clear that, for any given b, one can always determine the forcing frequency ω in such a way as to solve the complete determining equations and thus obtain a subharmonic solution of order 3. The approximate frequency-response curve is given as

$$\begin{aligned} \omega^2 &= 9\sigma^2\left[1 + \frac{3\epsilon\gamma}{4\sigma^2}(b^2 + bA_0 + 2A_0^2)\right] \\ A_0 &= -\frac{B}{8\sigma^2} \end{aligned} \tag{7-16}$$

This curve is also shown in Fig. 7-2 for $\epsilon\gamma > 0$ and $\epsilon\gamma < 0$. The maximum and minimum values of ω occur for $b = -A_0/2 = B/16\sigma^2$.

As an exercise discuss the cases where $p = 1$, $p = 3$. For $p = 1$, is it possible to have the forcing function in (7-12) independent of ϵ?

Now consider the Duffing equation with damping,

$$\begin{aligned} \dot{x}_1 &= x_2 \\ \dot{x}_2 &= -\sigma^2 x_1 - cx_2 + \epsilon\gamma x_1^3 + B \cos \omega t \end{aligned} \tag{7-17}$$

with damping coefficient c. Is it possible still to have subharmonic solutions of (7-17) even with $c \neq 0$? For $\sigma = 1/3$, Stoker [1] has shown

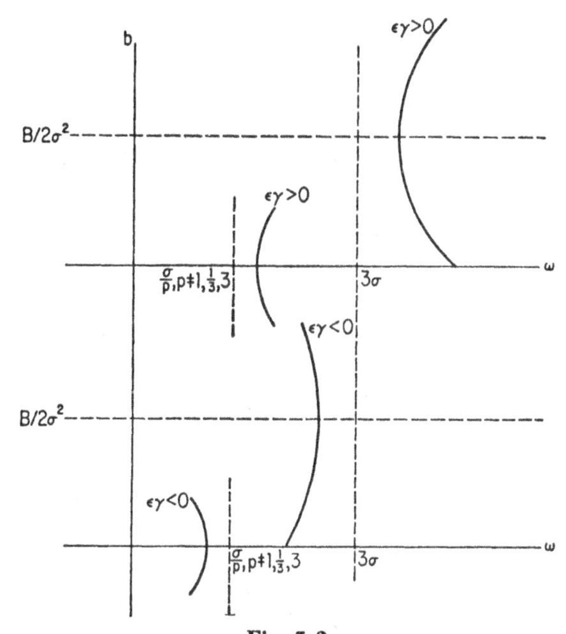

Fig. 7-2

that the damping coefficient c must be* $0(\epsilon)$ as $\epsilon \to 0$ in order to have a subharmonic of order 3. Prove this. Also, try to show that if $\sigma = 1/(2n + 1)$ then the damping coefficient must be $0(\epsilon^n)$ as $\epsilon \to 0$ in order to have a subharmonic or order $2n + 1$ (see Gambill and Hale [1] and Hale [2]).

Discuss the subharmonic solutions of the equation

$$\begin{aligned} \dot{x}_1 &= x_2 \\ \dot{x}_2 &= -\sigma^2 x_1 + \epsilon(3\nu \cos 2t - x_1^2) - \epsilon^2(\lambda x_2 + \mu x_1) - \epsilon^3 \mu x_1^2 \end{aligned}$$

* A function $c(\epsilon)$ is said to be $O(\epsilon)$ as $\epsilon \to 0$ if $c(\epsilon)/\epsilon$ is bounded as $\epsilon \to 0$. A function $c(\epsilon)$ is said to be $o(\epsilon)$ as $\epsilon \to 0$ if $c(\epsilon)/\epsilon \to 0$ as $\epsilon \to 0$.

where ϵ, ν, μ are parameters and ϵ is small. This equation has been discussed by Reuter [1] for the case $\sigma = 1$; that is, a subharmonic of order 2. To obtain a subharmonic of order n, show that it is necessary that λ be $0(\epsilon^{n-2})$.

For a good discussion of how subharmonic oscillations may be generated by mechanical devices, see Ludeke [1].

7-4. THE PROBLEM OF A CENTER

Consider the second-order system

$$\begin{aligned} \dot{x}_1 &= x_2 \\ \dot{x}_2 &= -\sigma^2 x_1 + \epsilon f(x_1,x_2) \end{aligned} \tag{7-18}$$

where $\sigma > 0$, ϵ are real parameters, $f(x_1,x_2)$ has continuous first and second derivatives with respect to x_1, x_2 for all x_1, x_2. The transformation

$$x_1 = \rho \sin \sigma\theta \qquad x_2 = \rho\sigma \cos \sigma\theta \tag{7-19}$$

leads to the system

$$\dot{\theta} = 1 - \frac{\epsilon}{\rho\sigma^2} f(\rho \sin \sigma\theta, \rho\sigma \cos \sigma\theta) \sin \sigma\theta$$

$$\dot{\rho} = \frac{\epsilon}{\sigma} f(\rho \sin \sigma\theta, \rho\sigma \cos \sigma\theta) \cos \sigma\theta$$

and thus

$$\frac{d\rho}{d\theta} = \frac{\epsilon}{\sigma} \frac{f(\rho \sin \sigma\theta, \rho\sigma \cos \sigma\theta) \cos \sigma\theta}{[1 - (\epsilon/\rho\sigma^2) f(\rho \sin \sigma\theta, \rho\sigma \cos \sigma\theta) \sin \sigma\theta]} \tag{7-20}$$

Theorem 7-1. If $f(0,0) = 0$ and either $f(x_1,-x_2) = f(x_1,x_2)$ or $f(-x_1,x_2) = -f(x_1,x_2)$, then for any constant a, $|a| \leq b$, there exists an $\epsilon_1 > 0$ and a function $\rho(\theta,a,\epsilon)$, $0 \leq |\epsilon| \leq \epsilon_1$, periodic in θ of period $2\pi/\sigma$ and satisfying (7-20); that is, system (7-18) has a center in the neighborhood of $x_1 = x_2 = 0$ (see Fig. 7-3).

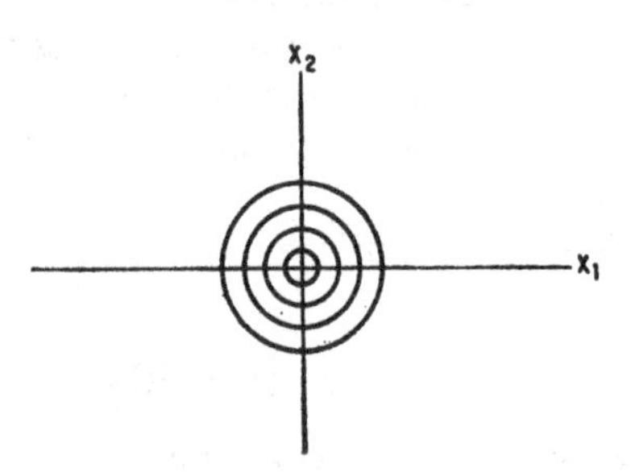

Fig. 7-3

Proof. Suppose first that $f(-x_1,x_2) = -f(x_1,x_2)$. Then equation (7-20) has property E (see Definition 6-1) with respect to $Q = (1)$ and the result follows from Corollary 6-1. If $f(x_1,-x_2) = f(x_1,x_2)$, then in (7-20) let $\theta = \varphi + \pi/\sigma$ and the new system again satisfies property E with respect to $Q = (1)$. The fact that system (7-18) has a center in the neighborhood of $x_1 = x_2 = 0$ follows from (7-19).

Equation (7-18) can be discussed in another way without introducing the polar-coordinate transformation (7-19). Also, by this method, the period of the solution will be exhibited in a more natural manner. By introducing the transformation

$$\begin{aligned} x_1 &= z_1 \sin \tau t + z_2 \cos \tau t \\ x_2 &= \tau(z_1 \cos \tau t - z_2 \sin \tau t) \end{aligned}$$

where $\tau^2 = \sigma^2 + \epsilon\beta$, we obtain

$$\begin{aligned} \dot{z}_1 &= \frac{\epsilon}{\tau} [\beta x_1 + f(x_1,x_2)] \cos \tau t \stackrel{\text{def}}{\equiv} \epsilon Z_1(t,z_1,z_2,\epsilon) \\ \dot{z}_2 &= -\frac{\epsilon}{\tau} [\beta x_1 + f(x_1,x_2)] \sin \tau t \stackrel{\text{def}}{\equiv} \epsilon Z_2(t,z_1,z_2,\epsilon) \end{aligned}$$

where x_1, x_2 are given by the expressions above. This system has property E with respect to $Q = \operatorname{diag}(1,-1)$ if $f(-x_1,x_2) = -f(x_1,x_2)$. Consequently, if we let $a^* = \operatorname{col}(a,0)$, then $Qa^* = a^*$ and if $z_1(t,a,\tau,\epsilon)$, $z_2(t,a,\tau,\epsilon)$ are the functions given by Theorem 6-1, then it follows from Theorem 6-5 that $Z_1(t,z_1(t,a,\tau,\epsilon),\ z_2(t,a,\tau,\epsilon))$ is an odd function of t. This means that the first determining equation in (6-14) is identically equal to zero for every constant a. Consequently, it is only necessary to investigate the second determining equation.

To the first approximation it is easily seen, from (7-2) with $z^{(0)} = \operatorname{col}(a,0)$, that the second determining equation is

$$\epsilon F(\beta,a) \stackrel{\text{def}}{\equiv} -\frac{\epsilon a}{2\sigma}\left(\beta + \frac{\sigma}{\pi a}\int_0^{2\pi/\sigma} f(a \sin \sigma t,\ a\sigma \cos \sigma t)(\sin \sigma t)\, dt\right) = 0$$

Since for $a \neq 0$, $F(\beta_0,a) = 0$ for

$$\beta_0 = -\frac{1}{\pi a}\int_0^{2\pi} f(a \sin \varphi,\ a\sigma \cos \varphi) \sin \varphi \, d\varphi$$

and $\partial F(\beta_0,a)/\partial\beta \neq 0$ for $a \neq 0$, the implicit-function theorem implies that the complete determining equations (6-14) have a solution $\beta(a,\epsilon)$, $\beta(a,0) = \beta_0$ for ϵ sufficiently small. This means there is a one-parameter family of periodic solutions of (7-18), the parameter being a. But, since (7-18) is autonomous, this implies a two-parameter family of solutions which is the same result as obtained before. The solution of (7-18) is given as

$$x_1[t, a, \tau(a,\epsilon), \epsilon] = z_1[t, a, \tau(a,\epsilon), \epsilon] \sin \tau(a,\epsilon)t + z_2[t, a, \tau(a,\epsilon), \epsilon] \cos \tau(a,\epsilon)t$$

$$x_2[t, a, \tau(a,\epsilon), \epsilon] = \frac{\partial x_1[t, a, \tau(a,\epsilon), \epsilon]}{\partial t}$$

where z_1, z_2 are given by Theorem 6-1 and $\tau^2(a,\epsilon) = \sigma^2 + \epsilon\beta(a,\epsilon)$ and $\beta(a,\epsilon)$ is the solution of the determining equations.

Let us discuss the first approximations of β and the maximum displacement $x_{1,\max}$ of the function $x_1[t, a, \tau(a,\epsilon), \epsilon]$ for the particular case where $f(x_1,x_2)$ in (7-19) is given by

$$f(x_1,x_2) = -\gamma x_1^3$$

that is, the approximation to the equation governing the motion of a simple pendulum for small oscillations.

For this simple case, $\beta_0 = 3\gamma a^2/4$ and the corresponding approximation to the frequency is

$$\tau^2 = \sigma^2 + \epsilon\beta_0 = \sigma^2 + \frac{3\epsilon\gamma a^2}{4}$$

If $\epsilon\gamma a^2 \neq 0$, then we can now observe a very interesting property of autonomous differential systems. The frequency of a periodic solution of an autonomous differential system in general varies with the parameters describing the differential equation. In particular, the frequency varies with ϵ.

The maximum displacement of $x_{1,\max}$ is approximately equal to a for ϵ small and thus we see that, for ϵ sufficiently small, the frequency τ varies with a. In fact, τ increases with a if $\gamma > 0$ (hard spring) and decreases with a if $\gamma < 0$ (soft spring). Schematically, the relationship between $x_{1,\max}$ and τ is shown in Fig. 7-4. These phenomena are of course well known and can be discussed without too much difficulty even for equations of this type without a small parameter (see, for example, Stoker [1]). However, we do not wish to discuss special techniques for two-dimensional systems but wish to emphasize that the methods we are using are designed for problems with many degrees of freedom and yield the proper kind of information even in dimension 2.

The case where $f(x_1,-x_2) = f(x_1,x_2)$ in (7-19) may be discussed in exactly the same way since the system for z_1, z_2 has property E with respect to $Q = (-1,1)$. System (7-18) with f even in x_2 has been discussed in much detail by McHarg [1] and Opial [1]. The interesting point to make about this system is that there may not be a two-parameter family of periodic solutions in every neighborhood of $x_1 = 0$, $x_2 = 0$ unless ϵ is small. In fact, if $\epsilon F(x_1,x_2)$ becomes too large, the phase portrait may be as in Fig. 7-5.

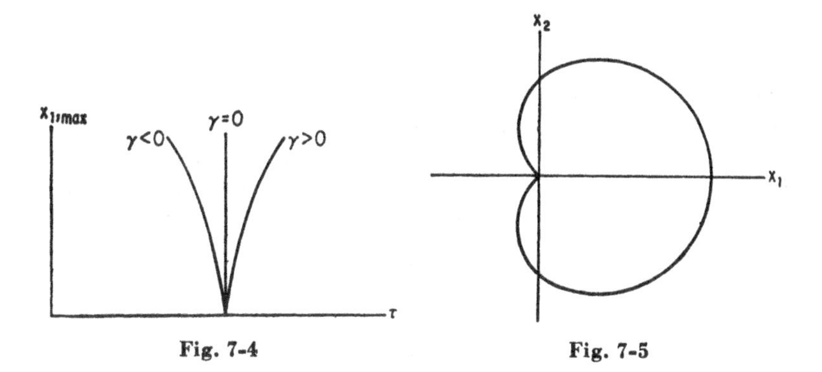

Fig. 7-4

Fig. 7-5

Extension of this example to higher-dimensional systems is given in Chap. 10.

7-5. AUTONOMOUS VAN DER POL EQUATION

Consider the equation

$$\begin{aligned} \dot{x}_1 &= x_2 \\ \dot{x}_2 &= -\sigma^2 x_1 + \epsilon(1 - x_1^2)x_2 \end{aligned} \tag{7-21}$$

where $\sigma > 0$, ϵ are real parameters and x_1, x_2 are scalars.

For nonlinear differential systems which do not contain the time (i.e., autonomous systems), we have seen in Sec. 7-4 that a periodic solution, if it exists, will have a period which, in general, varies with all the parameters in the systems, contrary to the situation in the non-autonomous systems. Consequently, to determine a periodic solution of (7-21), it is necessary to determine the period $2\pi/\tau$ as a function of ϵ as well as the amplitude. Of course, we want the frequency $\tau(\epsilon)$ to be such that $\tau(0) = \sigma$. This problem can be treated in many different ways, and we illustrate two of these.

The first approach is to use the transformation of van der Pol again, i.e., let

$$\begin{aligned} x_1 &= z_1 \sin \tau t + z_2 \cos \tau t \\ x_2 &= \tau(z_1 \cos \tau t - z_2 \sin \tau t) \end{aligned}$$

where $\tau^2 = \sigma^2 + \epsilon\beta$ is to be determined. The new system in z_1, z_2 is

$$\begin{aligned}\dot{z}_1 &= \frac{\epsilon}{\tau}[\beta x_1 + (1 - x_1^2)x_2]\cos \tau t \\ \dot{z}_2 &= -\frac{\epsilon}{\tau}[\beta x_1 + (1 - x_1^2)x_2]\sin \tau t\end{aligned} \tag{7-22}$$

where x_1, x_2 are given above and β, τ, σ are related by $\tau^2 = \sigma^2 + \epsilon\beta$. Since (7-22) is analytic in z_1, z_2, ϵ for all z_1, z_2 and ϵ sufficiently small, we may apply (7-1), (7-2). If a_1, a_2 are arbitrary real numbers and $z^{(0)} = \text{col}\,(a_1,a_2)$, then the first approximations to the determining equations (7-2) for $\epsilon \neq 0$ are easily seen to be equivalent to

$$\begin{aligned}F_1(a_1,a_2,\beta) &\overset{\text{def}}{\equiv} \beta a_2 + \sigma\left(1 - \frac{a_1^2 + a_2^2}{4}\right)a_1 = 0 \\ F_2(a_1,a_2,\beta) &\overset{\text{def}}{\equiv} \beta a_1 - \sigma\left(1 - \frac{a_1^2 + a_2^2}{4}\right)a_2 = 0\end{aligned}$$

Consequently, if we can find β_0, a_{10}, a_{20} such that $F_j(a_{10},a_{20},\beta_0) = 0$, $j = 1, 2$ and the rank of the matrix

$$\left[\frac{\partial F_j\,(a_{10},a_{20},\beta_0)}{\partial(a_1,a_2,\beta)}, j = 1,2\right]$$

is 2, then from the implicit-function theorem there will be a periodic solution of (7-22) close to col (a_{10},a_{20}) with a period close to $2\pi/(\sigma + \epsilon\beta_0)$ for ϵ sufficiently small. Take $a_{10}^2 + a_{20}^2 = 4$, $\beta_0 = 0$. Then

$$\left[\frac{\partial F_j\,(a_{10},a_{20},\beta_0)}{\partial(a_1,a_2,\beta)}, j = 1,2\right] = \begin{bmatrix} -\dfrac{\sigma a_{10}^2}{2} & -\dfrac{\sigma a_{20}a_{10}}{2} & a_{20} \\ +\dfrac{\sigma a_{20}a_{10}}{2} & \dfrac{\sigma a_{20}^2}{2} & a_{10}\end{bmatrix}$$

and the rank of this is equal to 2 for $a_{10}^2 + a_{20}^2 = 4$. Consequently, there is a periodic solution of (7-21) with amplitude equal to $2 + 0(\epsilon)$ and frequency equal to $\sigma + 0(\epsilon^2)$ as $\epsilon \to 0$.

Another manner of investigating this same problem is to introduce polar coordinates

$$x_1 = \rho \sin \sigma\theta \qquad x_2 = \rho\sigma \cos \sigma\theta$$

to obtain

$$\begin{aligned}\dot{\theta} &= 1 - \frac{\epsilon}{\sigma}(1 - \rho^2 \sin^2 \sigma\theta)\sin \sigma\theta \cos \sigma\theta \\ \dot{\rho} &= \epsilon(1 - \rho^2 \sin^2 \sigma\theta)\rho \cos^2 \sigma\theta\end{aligned}$$

For ϵ sufficiently small and ρ bounded, we can eliminate t in this equation to obtain

$$\frac{d\rho}{d\theta} = \epsilon \frac{(1 - \rho^2 \sin^2 \sigma\theta)\rho \cos^2 \sigma\theta}{1 - (\epsilon/\sigma)(1 - \rho^2 \sin^2 \sigma\theta) \sin \sigma\theta \cos \sigma\theta} \tag{7-23}$$

which is of the form (6-1) and $z = \rho$, $A = 0$, and the system is periodic in the independent variable θ of period $2\pi/\sigma$. Consequently, we can apply our theory and, in particular, (7-1), (7-2). For $k = 0$, $y^{(0)} = a$, a constant, the first approximations to the determining equations are

$$F(a) = a\left(1 - \frac{a^2}{4}\right) = 0$$

or $a = 0$, $a = \pm 2$. Obviously, $a = 0$ corresponds to the zero solution of (7-22). Since $F(2) = 0$, $F'(2) \neq 0$, it follows immediately from Theorem 6-4 that there is a solution $\rho(\theta,\epsilon)$ of (7-23), which is periodic in θ of period $2\pi/\sigma$ and $\rho(\theta,0) = 2$. To obtain a solution of the original equation (7-21) we need to know θ as a function of t. This is obtained by solving the equation

$$\dot{\theta} = 1 - \frac{\epsilon}{\sigma}[1 - \rho^2(\theta,\epsilon) \sin^2 \sigma\theta] \sin \sigma\theta \cos \sigma\theta$$

$$\theta(0) = \theta_0 \qquad \theta_0 \text{ arbitrary}$$

The period of the solution as a function of t is also obtained from this equation by solving for $\theta = \theta(t,\epsilon)$, $\theta(0,\epsilon) = 0$ and choosing $\tau = \tau(\epsilon)$ so that $\theta(\tau,\epsilon) = 2\pi/\sigma$. It follows that τ is a period in t since

$$\theta(t + \tau, \epsilon) = \theta(t,\epsilon) + 2\pi/\sigma$$

for all t.

7-6. MATHIEU'S EQUATION

Consider the Mathieu equation

$$\begin{aligned} \dot{x}_1 &= x_2 \\ \dot{x}_2 &= -\sigma^2 x_1 - \epsilon(\cos 2t)x_1 \end{aligned} \tag{7-24}$$

where $\sigma > 0$, ϵ are real parameters and x_1, x_2 are scalars. Let $\tau_1(\epsilon)$, $\tau_2(\epsilon)$, $\tau_1(0) = i\sigma$, $\tau_2(0) = -i\sigma$ be the characteristic exponents of system (7-24) and $\rho_j(\epsilon) = \exp \tau_j\pi$, $j = 1, 2$ be the characteristic multipliers. From the Floquet theory for linear equations with periodic coefficients,

$$\rho_1\rho_2 = 1 \qquad \text{for all } \epsilon$$

since the trace of the matrix associated with (7-24) is zero. Consequently, this implies that $(\tau_1 + \tau_2)\pi = 2k\pi i$ for some integer k. But since the characteristic exponents τ_j are determined only up to a multiple of 2, it follows that τ_1 may be taken equal to $-\tau_2$. [We have actually chosen a representation of τ_1, τ_2 which ensures this immediately for ϵ small since $\tau_1(0) = i\sigma$, $\tau_2(0) = -i\sigma$.] The characteristic multipliers of (7-24) satisfy the equation (since $\rho_1\rho_2 = 1$)

$$\rho^2 - 2A(\epsilon)\rho + 1 = 0$$

where $A(\epsilon)$ is a continuous function of ϵ at $\epsilon = 0$ and

$$A(0) = \tfrac{1}{2}[\rho_1(0) + \rho_2(0)] = \cos \pi\sigma$$

For the solutions of (7-24) to be bounded, it is necessary that $|\rho_j| = 1$, $j = 1, 2$, since $\rho_1\rho_2 = 1$. This latter condition is fulfilled if

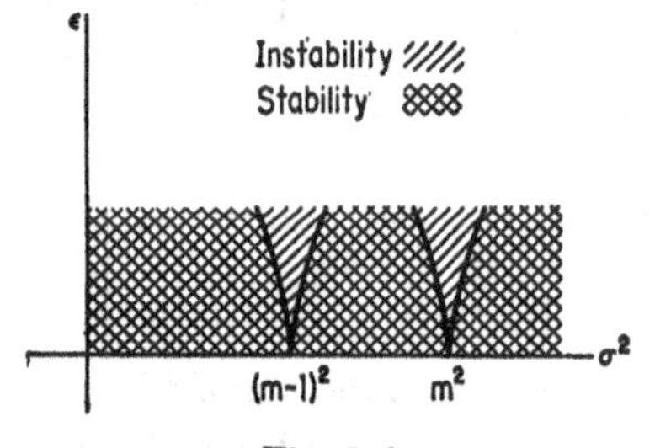

Fig. 7-6

and only if $|A(\epsilon)| \leq 1$. If $|A(\epsilon)| < 1$, then $\rho_1 \neq \rho_2$, $|\rho_j| = 1, j = 1, 2$, which implies boundedness of the solutions of (7-24) for all t in $(-\infty, \infty)$. Since $A(0) = \cos \pi\sigma$, it follows that, for every $\sigma \neq m$, $\sigma > 0$ there exists an $\epsilon_0 = \epsilon_0(\sigma)$ such that the solutions of (7-24) are bounded for $0 \leqq |\epsilon| \leqq \epsilon_0(\sigma)$, $-\infty < t < \infty$.

On the other hand, if $\sigma = m$, then $A(0) = \pm 1$ and one cannot conclude that the solutions are bounded by such a simple argument.

Suppose now that we consider σ as a parameter and try to determine σ as a function of ϵ in such a way that $A(\epsilon)$ is $+1$ (or -1) for all ϵ sufficiently small. Haupt [1] has shown that the curves obtained in the (σ,ϵ)-plane by this process divide the region of stability from the region of instability. Geometrically, the stability and instability regions are shown in Fig. 7-6. As one passes from a zone of stability to a zone of instability the characteristic multipliers vary from points on the unit circle in the complex plane to one multiplier inside the unit circle and one outside (see Fig. 7-7). This follows since $\rho_1\rho_2 = 1$ for all ϵ.

What we wish to do in this example is to show how the theory of Chap. 6 can be used to calculate for ϵ small the curves of transition from stability to instability.

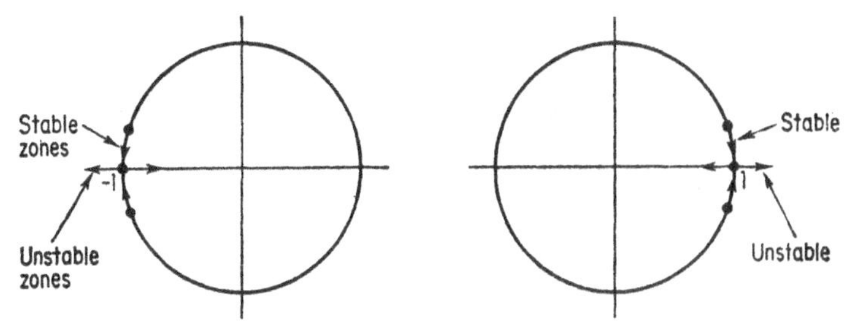

Fig. 7-7

For any given integer $m > 0$, the transformation

$$x_1 = z_1 \sin mt + z_2 \cos mt$$
$$x_2 = m(z_1 \cos mt - z_2 \sin mt)$$

applied to (7-24) yields

$$\begin{aligned} \dot{z}_1 &= \frac{1}{m}[(m^2 - \sigma^2) - \epsilon \cos 2t](z_1 \sin mt + z_2 \cos mt) \cos mt \\ \dot{z}_2 &= -\frac{1}{m}[(m^2 - \sigma^2) - \epsilon \cos 2t](z_1 \sin mt + z_2 \cos mt) \sin mt \end{aligned} \qquad (7\text{-}25)$$

What we do is determine σ as a function of ϵ, say $\sigma = \sigma(\epsilon)$, $\sigma(0) = m$, in such a way as to obtain a periodic solution of (7-25) [notice that the period of the coefficients of the right-hand side of (7-25) is π or 2π depending upon whether m is even or odd]. This periodic solution will yield one characteristic multiplier of (7-24) which is $+1$ if m is even and -1 if m is odd. Since the product of the multipliers is $+1$, the other multiplier must also be $+1$ if m is even and -1 if m is odd. The function $\sigma = \sigma(\epsilon)$ will be the desired transition curve.

Let us calculate these transition curves for $m = 1$ and 2. It is convenient to write (7-25) in vector form. If $z = \text{col}\,(z_1,z_2)$,

$$A = \begin{bmatrix} 1 & 0 \\ 0 & -1 \end{bmatrix} \qquad B = \begin{bmatrix} 0 & 1 \\ 1 & 0 \end{bmatrix} \qquad C = \begin{bmatrix} 0 & 1 \\ -1 & 0 \end{bmatrix} \qquad (7\text{-}26)$$

and $m^2 - \sigma^2 = \epsilon\beta$, then (7-25) becomes

$$\dot{z} = \frac{\epsilon}{2m}\left[\beta A \sin 2mt + \beta B \cos 2mt + \beta C - \frac{A}{2}\sin(2m+2)t - \frac{A}{2}\sin(2m-2)t - \frac{B}{2}\cos(2m+2)t - \frac{B}{2}\cos(2m-2)t - C\cos 2t\right]z \quad (7\text{-}27)$$

If $m = 1$ and we apply (7-2) with $z^{(0)} = \text{col}\,(a_1,a_2)$, then the first approximation to the determining equations is (except for a nonzero constant factor)

$$\left(\beta C - \frac{B}{2}\right)z^{(0)} = 0$$

or

$$\begin{aligned} a_2(\beta - \tfrac{1}{2}) &= 0 \\ a_1(\beta + \tfrac{1}{2}) &= 0 \end{aligned} \quad (7\text{-}28)$$

Since either a_1 or a_2 must be $\neq 0$ in order to obtain a nontrivial periodic solution, there are two possibilities for a solution of (7-28), namely, either $a_2 \neq 0$ arbitrary, $a_1 = 0$, $\beta = \frac{1}{2}$, or $a_1 \neq 0$ arbitrary, $a_2 = 0$, $\beta = -\frac{1}{2}$. The Jacobian matrix of the left-hand side of (7-28) is

$$\begin{bmatrix} 0 & \frac{1}{2}(\beta - \frac{1}{2}) & \dfrac{a_2}{2} \\ \frac{1}{2}(\beta + \frac{1}{2}) & 0 & \dfrac{a_1}{2} \end{bmatrix}$$

which has rank 2 at each of our solutions of (7-28). Consequently, from the implicit-function theorem there exists an exact solution of the determining equations close to the above approximate solutions for ϵ sufficiently small. Thus the transition curves in the (σ,ϵ)-plane in a neighborhood of the point (1,0) are to the first approximation given by $\sigma^2 = 1 \pm \frac{1}{2}\epsilon$.

Suppose $m \geq 2$. With $z^{(0)} = \text{col}\,(a_1,a_2)$, it follows from (7-1) that

$$z^{(1)} = z^{(0)} + \frac{\epsilon}{2m}\left[-\frac{\beta A}{2m}\cos 2mt + \frac{\beta B}{2m}\sin 2mt + \frac{A}{2(2m+2)}\cos(2m+2)t + \frac{A}{2(2m-2)}\cos(2m-2)t - \frac{B}{2(2m+2)}\sin(2m+2)t - \frac{B}{2(2m-2)}\sin(2m-2)t - \frac{C}{2}\sin 2t\right]z^{(0)} \quad (7\text{-}29)$$

If $m = 2$, it follows from (7-2) that the approximate determining equations up through terms of order ϵ are given by (except for a nonzero constant factor)

$$\left\{\beta C + \frac{\epsilon}{16}\left[\left(\beta^2 + \frac{2}{3}\right) C + B\right]\right\} z^{(0)} = 0$$

or, using (7-26),

$$\begin{aligned}\left[\beta + \frac{\epsilon}{16}\left(\beta^2 + \frac{5}{3}\right)\right] a_2 &= 0 \\ \left[\beta + \frac{\epsilon}{16}\left(\beta^2 - \frac{1}{3}\right)\right] a_1 &= 0\end{aligned} \tag{7-30}$$

Using the same type of argument as for the case of $m = 1$, one shows there are two distinct solutions of these equations, which are $\beta = \epsilon/48 + 0(\epsilon^2)$, $\beta = -5\epsilon/48 + 0(\epsilon^2)$ as $\epsilon \to 0$. The corresponding functions σ^2 obtained from $m^2 - \sigma^2 = \epsilon\beta$ are

$$\begin{aligned}\sigma^2 &= 4\left(1 - \frac{\epsilon^2}{192}\right) + 0(\epsilon^3) \\ &= 4\left(1 + \frac{5\epsilon^2}{192}\right) + 0(\epsilon^3)\end{aligned}$$

as $\epsilon \to 0$. Thus the two curves coincide up through the terms of order ϵ.

For $m > 2$, it follows from (7-2) and (7-29) that the approximate determining equations are (except for a nonzero constant factor)

$$\left[\beta - \frac{\epsilon}{2m}\left(\frac{\beta^2}{2m} + \frac{1}{4(2m+2)} + \frac{1}{4(2m-2)}\right)\right] Cz^{(0)} = 0 \tag{7-31}$$

On the other hand, these equations do not have two distinct solutions β as a function of ϵ. To obtain the transition curves for $m > 2$ one must therefore go to higher approximations $z^{(k)}$ until the two curves

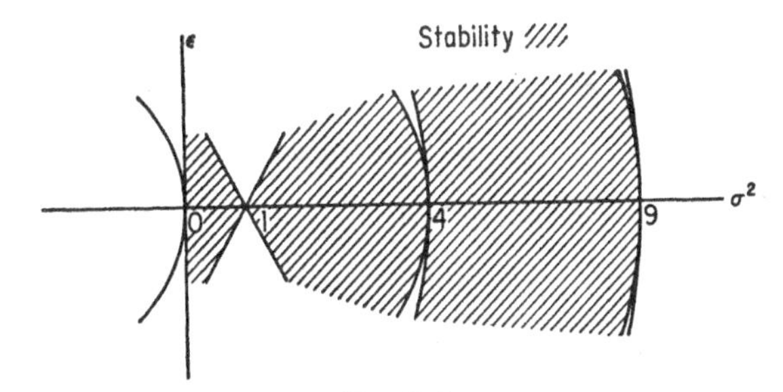

Fig. 7-8

$\sigma = \sigma(\epsilon)$ become different. Show that, for an arbitrary m, the transition curves coincide up through terms of ϵ^{m-1} and, in fact, must be given as $\sigma^2 = m^2 + O(\epsilon^m)$ as $\epsilon \to 0$. This last result was proved by Hale [3], and in that paper, the same type of result was obtained for even higher-order systems of Mathieu-type equations. Figure 7-8 gives the approximate nature of the transition curves for (7-24) near $\epsilon = 0$. The shaded regions correspond to stability. We have not discussed the case $m = 0$, but this will be done in Sec. 7-8.

7-7. STABILITY OF PERIODIC SOLUTIONS

Consider the second-order system

$$\begin{aligned} \dot{w}_1 &= w_2 \\ \dot{w}_2 &= -\omega^2 w_1 + \epsilon f(t,w_1,w_2) \end{aligned} \tag{7-32}$$

where $\omega > 0$, ϵ are real numbers, $f(t,w_1,w_2)$ is periodic in t of period $T = 2\pi/\omega$, is continuous in t, w_1, w_2, and has continuous first and second derivatives with respect to w_1, w_2 for all t, w_1, w_2. We further suppose that (7-32) has a periodic solution $w_1^0(t,\epsilon)$, $w_2^0(t,\epsilon)$ of period T, continuous and continuously differentiable with respect to ϵ at $\epsilon = 0$. Our problem is to find conditions on the function $f(t,w_1,w_2)$ which will ensure that this periodic solution is asymptotically stable in the sense of Liapunov.

If $w_1 = w_1^0 + x_1$, $w_2 = w_2^0 + x_2$, then the linear approximation to the equation for x_1, x_2 is

$$\begin{aligned} \dot{x}_1 &= x_2 \\ \dot{x}_2 &= -\omega^2 x_1 + \epsilon a(t)x_1 + \epsilon b(t)x_2 \end{aligned} \tag{7-33}$$

where

$$\begin{aligned} a(t) &= \partial f\,(t,w_1^0(t,\epsilon),w_2^0(t,\epsilon))/\partial w_1 \\ b(t) &= \partial f\,(t,w_1^0(t,\epsilon),w_2^0(t,\epsilon))/\partial w_2 \end{aligned} \tag{7-34}$$

With our usual periodic transformation (7-4) applied to (7-33), we have

$$\begin{aligned} \dot{z}_1 &= \frac{\epsilon}{\omega}\,[a(t)(z_1 \sin \omega t + z_2 \cos \omega t) \\ &\qquad\qquad + \omega b(t)(z_1 \cos \omega t - z_2 \sin \omega t)] \cos \omega t \\ \dot{z}_2 &= -\frac{\epsilon}{\omega}\,[a(t)(z_1 \sin \omega t + z_2 \cos \omega t) \\ &\qquad\qquad + \omega b(t)(z_1 \cos \omega t - z_2 \sin \omega t)] \sin \omega t \end{aligned} \tag{7-35}$$

The theorem of Liapunov assures us that, if the solutions of (7-33) approach zero as $t \to \infty$ or, equivalently, if the characteristic exponents of (7-35) have negative real parts, then the original periodic solution of

(7-32) is asymptotically stable. Consequently, we need only obtain the characteristic exponents of system (7-35). To each characteristic exponent τ of (7-35), it is known that there is at least one solution of the form $z_1 = e^{\tau t}p_1(t)$, $z_2 = e^{\tau t}p_2(t)$ where p_1, p_2 are periodic of period $T = 2\pi/\omega$. Consequently, if we let $z_1 = e^{\tau t}p_1$, $z_2 = e^{\tau t}p_2$, the equations for p_1, p_2 are

$$\begin{aligned} \dot{p}_1 &= \left[-\tau + \frac{\epsilon}{2}b(t) + \frac{\epsilon}{2\omega}a(t)\sin 2\omega t + \frac{\epsilon}{2}b(t)\cos 2\omega t\right]p_1 \\ &\qquad + \left[\frac{\epsilon}{2\omega}a(t) + \frac{\epsilon}{2\omega}a(t)\cos 2\omega t - \frac{\epsilon}{2}b(t)\sin 2\omega t\right]p_2 \\ \dot{p}_2 &= \left[-\frac{\epsilon}{2\omega}a(t) + \frac{\epsilon}{2\omega}a(t)\cos 2\omega t - \frac{\epsilon}{2}b(t)\sin 2\omega t\right]p_1 \\ &\qquad + \left[-\tau + \frac{\epsilon}{2}b(t) - \frac{\epsilon}{2\omega}a(t)\sin 2\omega t - \frac{\epsilon}{2}b(t)\cos 2\omega t\right]p_2 \end{aligned} \tag{7-36}$$

Our problem is now to determine τ in such a way that this equation (7-36) has a periodic solution. For $\epsilon = 0$, τ must be zero. If we let $\tau = \epsilon\beta$, then system (7-36) becomes a special case of system (6-1) with $z = \text{col}\,(p_1,p_2)$ and A equal to the zero matrix. If a_1, a_2 are arbitrary numbers, then it is easy to see by applying (6-9) that the determining equations are linear in a_1, a_2. Furthermore, the first approximations to these equations are

$$\begin{bmatrix} c_{11} - \beta & c_{12} \\ c_{21} & c_{22} - \beta \end{bmatrix} \begin{bmatrix} a_1 \\ a_2 \end{bmatrix} = 0 \tag{7-37}$$

where $c_{11} - \beta$, c_{12}, c_{21}, $c_{22} - \beta$ are the mean values of the obvious coefficients in (7-36). Consequently, we can state the following result: If the eigenvalues of the matrix (c_{ij}), $i, j = 1, 2$ have negative real parts, then the periodic solution of (7-32) is asymptotically stable. One can easily write out explicitly the meaning of this in terms of the functions $a(t)$, $b(t)$ in (7-34), as was done by Mandelstam and Papalexi [1]. With this criterion for stability, is the periodic solution in Sec. 7-1 asymptotically stable? What about the stability of the periodic solution of (7-23)? Notice the difference between the stability of the periodic solution obtained in Sec. 7-1 and the one in Sec. 7-5.

7-8. A CASE OF NONSIMPLE ELEMENTARY DIVISORS

Consider the system

$$\begin{aligned} \dot{x}_1 &= x_2 \\ \dot{x}_2 &= \mu f(t, x_1, x_2) \end{aligned} \tag{7-38}$$

where $\mu > 0$ is a real number and $f(t,x_1,x_2)$ is periodic in t of period $2\pi/\omega$, is continuous in t, x_1, x_2 and analytic in x_1, x_2 in some region. This system is of a different nature from the previous examples since, for $\mu = 0$, all the solutions of (7-38) are not periodic even though both the characteristic roots are zero; in fact, some solutions are unbounded. But we can still phrase the question: For μ sufficiently small, does there exist a periodic solution of (7-38) which reduces to a periodic solution of (7-38) for $\mu = 0$, namely, to $x_1 = $ constant, $x_2 = 0$?

The easiest manner in which to answer this question is to let $\epsilon = \mu^{1/2}$, $x_1 = z_1$, $x_2 = \epsilon z_2$ to obtain the new system

$$\begin{aligned} \dot{z}_1 &= \epsilon z_2 \\ \dot{z}_2 &= \epsilon f(t,z_1,\epsilon z_2) \end{aligned} \tag{7-39}$$

which is of the form (6-1) with $z_1 = $ col (z_1,z_2) and A equal to the zero matrix.

We may apply the procedure (6-9). If a_1, a_2 are arbitrary and $z^{(0)} = (a_1,a_2)$, then the first approximation to the determining equations (6-11) or (6-14) is

$$\begin{aligned} a_2 &= 0 \\ \frac{1}{T}\int_0^T f(t,a_1,0)\,dt &= 0 \end{aligned} \tag{7-40}$$

If the second equation in (7-40) has a simple root a_{10}, then there will be a periodic solution of (7-38). Since $f(t,x_1,x_2)$ is analytic in x_1, x_2, this solution is analytic in $\mu^{1/2}$ at $\mu = 0$, and for $\mu = 0$, this solution is $x_1 = a_{10}$, $x_2 = 0$. Let us consider a special case of (7-38), namely, the nonlinear Mathieu equation with a small basic frequency

$$f(t,x_1,x_2) = (\sigma + \alpha \cos 2t)x_1 + bx_1^3$$

For this f, the second equation in (7-40) becomes

$$(\sigma + ba_1^2)a_1 = 0$$

which has the simple solutions $a_1 = 0$, $a_1 = \pm(-\sigma/b)^{1/2}$ if $b\sigma < 0$. The solution $a_1 = 0$ clearly corresponds to the zero solution of our equation and the other solutions correspond to periodic solutions of period π.

As another example, consider the determination of the transition curve near $m = 0$ from a stability region in the (σ,ϵ)-plane to an instability region in the (σ,ϵ)-plane of the Mathieu equation (7-24) of Sec. 7-6.

If $\epsilon > 0$ in (7-24), $x_1 = z_1$, $x_2 = \sqrt{\epsilon}\, z_2$, then (7-24) becomes

$$\begin{aligned} \dot{z}_1 &= \sqrt{\epsilon}\, z_2 \\ \dot{z}_2 &= -\frac{\sigma^2}{\sqrt{\epsilon}} z_1 - \sqrt{\epsilon}\,(\cos 2t) z_1 \end{aligned} \tag{7-41}$$

If $\sigma^2 = \epsilon\beta$, where β is to be determined, then (7-41) becomes

$$\begin{aligned} \dot{z}_1 &= \sqrt{\epsilon}\, z_2 \\ \dot{z}_2 &= -\sqrt{\epsilon}\,(\beta + \cos 2t) z_1 \end{aligned} \tag{7-42}$$

which is of the form (7-39) with ϵ replaced by $\sqrt{\epsilon}$. We now try to determine β in such a way as to obtain a periodic solution of period π of (7-42). Proceeding with the iteration procedure (7-1) with $z^{(0)} = \text{col}\,(a_1, a_2)$ we obtain

$$\begin{cases} z^{(1)} = \text{col}\,(z_1^{(1)}, z_2^{(1)}) \\ z_1^{(1)} = a_1 \\ z_2^{(1)} = a_2 - \dfrac{\sqrt{\epsilon}}{2}(\sin 2t) a_1 \end{cases}$$

$$\begin{cases} z^{(2)} = \text{col}\,(z_1^{(2)}, z_2^{(2)}) \\ z_1^{(2)} = a_1 + \dfrac{\epsilon}{4}(\cos 2t) a_1 \\ z_2^{(2)} = a_2 - \dfrac{\sqrt{\epsilon}}{2}(\sin 2t) a_1 \end{cases}$$

Consequently, with $z^{(2)}$ as above, it follows that the approximate determining equations are given by (except for a nonzero constant factor)

$$\begin{aligned} a_2 &= 0 \\ \left(\beta + \frac{\epsilon}{8}\right) a_1 &= 0 \end{aligned} \tag{7-43}$$

These equations have the simple solution $\beta = -\epsilon/8$, $a_2 = 0$, and a_1 arbitrary. Consequently, the complete determining equations have a solution and, since $\sigma^2 = \epsilon\beta$, the first approximation to the transition curve at $m = 0$ is given by $\sigma^2 = -\epsilon^2/8$. This curve is shown in Fig. 7-8. Notice that σ^2 must be less than zero and thus the stability region lies on the left-hand side of the ϵ axis.

8: Characteristic Exponents of Linear Periodic Systems

The purpose of the present chapter is to show how the problem of finding characteristic exponents of a general class of linear periodic differential systems may be reduced to the problem of determining periodic solutions of a nonautonomous system in the standard form (6-1). The determining equations are discussed in detail. Also, some of the properties of reciprocal systems are given.

Consider the system of equations

$$\dot{w} = Cw + \epsilon\Phi(t,\epsilon)w \tag{8-1}$$

where ϵ is a real parameter, $0 \leqq \epsilon \leqq \epsilon_0$, w is an n vector, C is an $n \times n$ real constant matrix, $\Phi(t,\epsilon)$ is an $n \times n$ real matrix whose elements are periodic in t of period $T = 2\pi/\omega$, $\omega > 0$, L integrable in $[0,T]$, having continuous second derivatives with respect to ϵ, and there exists a function $\eta(t)$, L integrable in $[0,T]$, such that $\|\Phi(t,\epsilon)\| \leqq \eta(t)$, $0 \leqq t \leqq T, 0 \leqq \epsilon \leqq \epsilon_0$. We also suppose that $C = \text{diag}\,(B_1,B_2)$ where B_1 is a $p \times p$ matrix whose eigenvalues have simple elementary divisors and all the eigenvalues of $\exp\,(B_1T)$ are equal to the same number ρ_0, while no eigenvalue of $\exp\,(B_2T)$ is equal to ρ_0. We shall briefly refer to system (8-1) satisfying all the above conditions as *system* (8-1).*

If the eigenvalues of the matrix C are $\lambda_1, \ldots, \lambda_n$, then the above assumptions imply that

$$\begin{aligned} &\lambda_j - \lambda_k \equiv 0 \ (\text{mod}\ \omega i) \qquad && j, k = 1, 2, \ldots, p \\ &\lambda_l - \lambda_j \not\equiv 0 \ (\text{mod}\ \omega i) \qquad && j = 1, 2, \ldots, p \\ & && l = p+1, \ldots, n \end{aligned} \tag{8-2}$$

* By a solution of system (8-1), we shall mean a continuous function which has a derivative almost everywhere and satisfies (8-1) almost everywhere.

Under the above conditions, the characteristic multipliers of system (8-1) are continuous functions of ϵ at $\epsilon = 0$. We wish to determine the characteristic multipliers $\rho_j(\epsilon)$, $j = 1, 2, \ldots, p$, for which $\rho_j(0) = \rho_0$, $j = 1, 2, \ldots, p$. If we define $\rho_0 = \exp(\lambda_1 T)$, then the characteristic exponents $\tau_j(\epsilon)$ defined by $\rho_j(\epsilon) = \exp(\tau_j(\epsilon)T)$, $j = 1, 2, \ldots, p$, can also be taken as continuous functions of ϵ at $\epsilon = 0$ if we define $\tau_j(0) = \lambda_1$, $j = 1, 2, \ldots, p$. Furthermore, since (8-1) is a real system, it follows that $\bar{\tau}$, the complex conjugate of τ, is a characteristic exponent if τ is.

Suppose that a transformation of variables has been made in (8-1) so that B_1 is in diagonal form. Let β be any complex number $|\beta| \leqq R$, and choose $\epsilon_0(R)$ so small that $\lambda_l - \lambda_1 - \epsilon\beta \not\equiv 0 \pmod{\omega i}$, $l = p + 1, \ldots, n$, $0 \leqq \epsilon \leqq \epsilon_0$; let $w = \text{col}(u,v)$, where u is a p vector and consider the transformation

$$\begin{aligned} u &= e^{(\lambda_1+\epsilon\beta)t}e^{(B_1-\lambda_1 I)t}x \\ v &= e^{(\lambda_1+\epsilon\beta)t}y \end{aligned} \tag{8-3}$$

where x is a p vector and I will always denote the identity matrix of appropriate dimension. Notice that the matrix $\exp(B_1 - \lambda_1 I)t$ is periodic of period T. If we partition the matrix Φ as $\Phi = (\Phi_{jk})$, $j, k = 1, 2$, where Φ_{11} is a $p \times p$ matrix, then the differential equations for $z = (x,y)$ are

$$\begin{aligned} \dot{x} &= -\epsilon\beta x + \epsilon e^{-(B_1-\lambda_1 I)t}\Phi_{11}(t,\epsilon)e^{(B_1-\lambda_1 I)t}x + \epsilon e^{-(B_1-\lambda_1 I)t}\Phi_{12}(t,\epsilon)y \\ \dot{y} &= [B_2 - (\lambda_1 + \epsilon\beta)I]y + \epsilon\Phi_{21}(t,\epsilon)e^{(B_1-\lambda_1 I)t}x + \epsilon\Phi_{22}(t,\epsilon)y \end{aligned} \tag{8-4}$$

From assumption (8-2) on the λ_j, this system can be written in the form

$$\dot{z} = Bz + \epsilon\psi(t,\beta,\epsilon)z \tag{8-5}$$

where $\psi(t,\beta,\epsilon)$ satisfies the same conditions as the matrix $\Phi(t,\epsilon)$ in (8-1), $z = \text{col}(x,y)$, $B = \text{diag}(0, B_2 - \lambda_1 I)$, and 0 is the zero matrix of appropriate dimension.

If we can determine β in such a way that (8-5) has a periodic solution of period T, then this solution yields, from (8-3), a solution of (8-1) of the form

$$w = e^{(\lambda_1+\epsilon\beta)t}p(t) \qquad p(t + T) = p(t) \tag{8-6}$$

which implies that $\lambda_1 + \epsilon\beta$ is a characteristic exponent of (8-1).

The determination of periodic solutions of system (8-5) is a special case of the results of Chap. 6. In fact, for any given constant p vector

a, Theorem 6-1 asserts the existence of a function

$$z(t,a,\beta,\epsilon) = \text{col}\,(x(t,a,\beta,\epsilon),\, y(t,a,\beta,\epsilon)),\; z(t,a,\beta,0) = \text{col}\,(a,0)$$

which is periodic in t of period T and satisfies the equation

$$\dot{z} = Bz + \epsilon\psi(t,\beta,\epsilon)z - \epsilon F^*(a,\beta,\epsilon) \tag{8-7}$$

where $F^* = \text{col}\,(F,0)$, F a p vector, and

$$TF(a,\beta,\epsilon) = \int_0^T e^{-(B_1-\lambda_1 I)t}[\Phi_{11}(t,\epsilon)e^{(B_1-\lambda_1 I)t}x(t,a,\beta,\epsilon) + \Phi_{12}(t,\epsilon)y(t,a,\beta,\epsilon)]\,dt - \beta Ta \tag{8-8}$$

One can show as in the proof of Theorem 6-1 that $F(a,\beta,\epsilon)$ has continuous second derivatives with respect to ϵ, all derivatives with respect to β, and furthermore,

$$F(a,\beta,\epsilon) = [G(\beta,\epsilon) - \beta I]a \tag{8-9}$$

where

$$G(\beta,0) = \frac{1}{T}\int_0^T e^{-(B_1-\lambda_1 I)t}\Phi_{11}(t,0)e^{(B_1-\lambda_1 I)t}\,dt \tag{8-10}$$

a matrix which is independent of β.

Theorem 8-1. Let $G(\beta,\epsilon)$, $G(\beta,0)$ be defined by (8-8), (8-9), (8-10) and let β_j^0, $j = 1, 2, \ldots, p$, be the eigenvalues of the matrix $G(\beta,0)$. Then for ϵ sufficiently small, the p characteristic exponents of system (8-1) which are $\equiv \lambda_1 \pmod{\omega i}$ for $\epsilon = 0$ are of the form $\lambda_1 + \epsilon\beta_j(\epsilon)$, where the $\beta_j(\epsilon)$, $j = 1, 2, \ldots, p$, are solutions of the equation

$$\det\,[G(\beta,\epsilon) - \beta I] = 0 \tag{8-11}$$

with $\beta_j(0) = \beta_j^0$, $j = 1, 2, \ldots, p$. The complex conjugates of $\lambda_1 + \epsilon\beta_j(\epsilon)$, $j = 1, 2, \ldots, p$, are also characteristic exponents of system (8-1).

Proof. Since $G(\beta,0)$ is independent of β, equation (8-11) for $\epsilon = 0$ is a polynomial in β of degree p. By a simple argument from index theory it follows that, for ϵ sufficiently small, equation (8-11) has p solutions of the above type. But, Theorem 6-2 implies that a necessary and sufficient condition that equations (8-5) have a periodic solution, for ϵ sufficiently small, is that β, a are such that $F(a,\beta,\epsilon) = 0$, where $F(a,\beta,\epsilon)$ is defined in (8-8), and this last relation can hold if and only if β satisfies (8-11). Finally, $\lambda_1 + \epsilon\beta$ is a characteristic exponent if and only if (8-5) has a periodic solution continuous in ϵ, and the theorem follows immediately.

Corollary 8-1. If λ_1 is purely imaginary and the eigenvalues of the matrix $G(\beta,0)$ in (8-10) have negative real parts, then there exist p linearly independent solutions of (8-1) which approach zero as $t \to \infty$. If $\lambda_1 \not\equiv 0 \pmod{\omega i}$, then there are $2p$ such solutions.

If λ_1 is not purely imaginary then, for ϵ sufficiently small, the corresponding characteristic exponents will have the same real parts as λ_1 and no method of successive approximations is necessary to determine the qualitative behavior of the solutions. This remark, together with Corollary 8-1, is the basis for most of the criteria for stability of periodic solutions of weakly nonlinear differential equations. See, for example, Bailey and Gambill [1], Blehman [1,2], Hale [4], Nohel [1], Sibuya [2]. If one has some information beforehand to the effect that one or more of the characteristic exponents is zero (as in studying the stability of periodic solutions of weakly nonlinear autonomous systems) then this information can be used very efficiently in (8-11). See, for example, Hale [4]. These remarks are illustrated in Chaps. 8 and 9 by proving some stability theorems for periodic solutions of nonlinear equations. Another derivation of these results is given in Chap. 17.

In case $p = 1$, i.e., the function $F(a,\beta,\epsilon)$ in (8-8) is a scalar function, then β (and, therefore, the characteristic exponent) can be calculated very easily by successive approximations since the function $y(t,a,\beta,\epsilon)$ in (8-8) is determined by the iterative scheme (6-9). From (8-9), (8-10), it follows that for $p = 1$ the first approximation to the characteristic exponent close to λ_1 is $\lambda_1 + \epsilon\beta_0$ where β_0 is the mean value of the scalar function $\Phi_{11}(t,0)$. Even in case $p > 1$, it is not too difficult to determine successive terms of $F(a,\beta,\epsilon)$ in (8-8), but for p large, the solution of the equation (8-11) is, of course, difficult.

Corollary 8-2. If $\Phi(t,\epsilon)$ is analytic in ϵ, $0 \leqq \epsilon \leqq \epsilon_0$, and β_0 is an eigenvalue of $G(\beta,0)$ of multiplicity r, then there exist r solutions $\beta_j(\epsilon)$, $j = 1, 2, \ldots, r$, of equation (8-11) which are analytic in $\epsilon^{1/q}$ for some integer q and have the form

$$\beta_j(\epsilon) = \beta_0 + \sum_{k=1}^{\infty} \alpha_{jk}\epsilon^{k/q} \qquad j = 1, 2, \ldots, r \tag{8-12}$$

Proof. It is easy to show that $\Phi(t,\epsilon)$ analytic in ϵ implies that $G(\beta,\epsilon)$ is analytic in ϵ and β. The result then follows from the well-known results in the theory of complex variables.

Corollary 8-1 and the remarks following that result are useful for obtaining theorems concerning asymptotic stability of the solutions of (8-1) on $(0, \infty)$. To obtain boundedness theorems on $(-\infty, \infty)$ one must discuss the qualitative structure of the functions $F(a,\beta,\epsilon)$ in (8-8). However, we choose to proceed in another direction taking advantage of results that can be obtained without using successive approximations.

Consider a more general system of the form

$$\dot{z} = P(t,\epsilon)z \tag{8-13}$$

where $P(t,\epsilon)$ is a real function and satisfies the same conditions as $\Phi(t,\epsilon)$ in (8-1). Following Liapunov [1], system (8-13) will be said to be *reciprocal* if for every *characteristic multiplier* ρ of (8-13) there is also the characteristic multiplier ρ^{-1}. The next three theorems are essentially due to Liapunov [1].

Theorem 8-2. If system (8-13) is reciprocal and ρ_0, $|\rho_0| = 1$, is a simple characteristic multiplier of (8-13) for $\epsilon = 0$, then there exists an ϵ_1, $0 < \epsilon_1 \leqq \epsilon_0$, such that there is a characteristic multiplier $\rho(\epsilon)$ of (8-13) with $|\rho(\epsilon)| = 1$, $\rho(0) = \rho_0$, $0 \leqq |\epsilon| \leqq \epsilon_1$.

Proof. If ρ_0 is a simple characteristic multiplier of (8-13) for $\epsilon = 0$, then there exists an ϵ_1 such that the characteristic multiplier $\rho(\epsilon)$ of (8-13) with $\rho(0) = \rho_0$ is simple for $0 \leqq |\epsilon| \leqq \epsilon_1$. Since system (8-13) is assumed to be reciprocal, it follows that $\rho^{-1}(\epsilon)$ is also a characteristic multiplier. System (8-13) is a real system and so $\bar{\rho}(\epsilon)$ is also a simple multiplier. The conclusion of the theorem is now obvious.

Theorem 8-3. If system (8-13) is reciprocal and all characteristic multipliers are distinct and have unit modulus for $\epsilon = 0$, then there exists an ϵ_1, $0 < \epsilon_1 \leqq \epsilon_0$, such that all solutions of (8-13) are bounded in $-\infty < t < +\infty$, $0 \leqq |\epsilon| \leqq \epsilon_1$.

Proof. Theorem 8-2 implies all the characteristic multipliers of (8-13) in this case are simple and on the unit circle for $0 \leqq |\epsilon| \leqq \epsilon_1$, $\epsilon_1 > 0$. The result is then immediate from the Floquet representation in Theorem 3-1.

Theorem 8-4. If there exists a matrix $B(\epsilon)$ of order n, continuous in ϵ at $\epsilon = 0$, $|\det B(\epsilon)| \geqq \delta > 0$, $0 \leqq |\epsilon| \leq \epsilon_0$, such that either of the following conditions is satisfied,

$$B(\epsilon)P(t,\epsilon) = -P(-t,\epsilon)B(\epsilon) \qquad 0 \leqq |\epsilon| \leqq \epsilon_0 \qquad -\infty < t < +\infty \tag{a}$$

$$B(\epsilon)P(t,\epsilon) = -P'(t,\epsilon)B(\epsilon) \qquad 0 \leqq |\epsilon| \leqq \epsilon_0 \qquad -\infty < t < +\infty \tag{b}$$

(P' is the transpose of P)

then system (8-13) is reciprocal.

Proof. Let $Z(t)$, $Z(0) = I$, be a fundamental system of real solutions of (8-13). If $Y(t)$, $Y(t) = Y_0$ is a fundamental system of solutions of the adjoint equation of (8-13),

$$\dot{y} = -yP(t,\epsilon)$$

then an easy calculation shows that

$$Y(t)Z(t) = Y_0$$

for all t.

1. If $B(\epsilon)P(t,\epsilon)B^{-1}(\epsilon) = -P(-t,\epsilon)$ then $Y(t) = Z^{-1}(-t)B(\epsilon)$ satisfies the adjoint equation of (8-13). Therefore,

$$Z^{-1}(-t)B(\epsilon)Z(t) = B^{-1}(\epsilon)$$

or $Z(t) = B^{-1}(\epsilon)Z(-t)B(\epsilon)$. If the Floquet representation of $Z(t)$ is $Z(t) = Q(t)e^{tA}$, then for each eigenvalue ρ of the equation

$$\det[e^{TA} - \rho I] = 0$$

it follows that $\det[\rho^{-1}I - e^{TA}] = 0$; that is, the characteristic equation is reciprocal.

2. If $B(\epsilon)P(t,\epsilon)B^{-1}(\epsilon) = -P'(t,\epsilon)$, then $Y(t) = Z'(t)B(\epsilon)$ is a solution of the adjoint equation of (8-13) and $Z'(t)B(\epsilon)Z(t) = B(\epsilon)$, or $Z'(t) = B(\epsilon)Z^{-1}(t)B^{-1}(\epsilon)$. Consequently, the characteristic equation $\det[Z(T) - \rho I] = 0$ has the same roots as the equation

$$\det[Z^{-1}(T) - \rho I] = 0$$

and the result follows immediately.

Let us now return to systems of the form (8-1) and, in particular, consider a system of second-order equations

$$\ddot{w} + Aw = \epsilon\Phi(t)w \tag{8-14}$$

where $A = \text{diag}(\sigma_1^2, \ldots, \sigma_n^2)$, $\sigma_j > 0$, $\Phi(t)$ is periodic in t of period $T = 2\pi/\omega$, L integrable in $[0,T]$. If one transforms system (8-14) by the standard procedure to a system of first-order equations, then one can show by Theorem 8-4 that (8-14) is reciprocal in either of the following cases:

1. $\Phi(t) = (\Phi_{jk}(t))$
where each $\Phi_{jk}(t)$ is a matrix and $\Phi_{jk}(-t) = (-1)^{k+j}\Phi_{jk}(t)$.
2. $\Phi(t)$ is symmetric.

Consequently, from Theorem 8-3, all solutions of (8-14) will be bounded in $(-\infty,+\infty)$ for ϵ sufficiently small if the matrix $\Phi(t)$

satisfies either condition 1 or 2 and the numbers σ_j satisfy the relations

$$2\sigma_j \not\equiv 0 \qquad \sigma_j \pm \sigma_k \not\equiv 0 \pmod{\omega} \qquad j \neq k \qquad j, k = 1, 2, \ldots, n \tag{8-15}$$

Condition (8-15) implies that the characteristic multipliers of (8-14) are distinct for $\epsilon = 0$.

Now, the natural question to ask is whether or not it is possible that some of the conditions (8-15) may not be satisfied and the solutions of (8-14) are still bounded for ϵ sufficiently small.

Gel'fand and Lidskii [1] and, independently, Moser [1] and Yakubovich [4] have shown that, if $\Phi(t)$ satisfies condition 2, then the solutions of (8-14) are bounded if (8-15) is replaced by the weaker hypotheses

$$\sigma_j + \sigma_k \not\equiv 0 \pmod{\omega} \qquad j, k = 1, 2, \ldots, n \tag{8-16}$$

The analysis given by the above authors does not depend on successive approximations but rather on the algebraic properties of a fundamental system of solutions, namely, that this fundamental system belongs to the class of symplectic matrices.

Unfortunately, if $\Phi(t)$ satisfies condition 1, it does not seem possible to obtain such beautiful results. Also, in case some of the conditions in (8-16) are not satisfied, one needs some procedure for investigating the behavior of the characteristic exponents. The following result gives a procedure for investigating the case of reciprocal systems with multiple characteristic multipliers for $\epsilon = 0$.

Theorem 8-5. Suppose that system (8-1) is real, analytic in ϵ, reciprocal, and that the eigenvalue λ_1 is purely imaginary and condition (8-2) is satisfied. Suppose β_0 is an eigenvalue of $G(\beta,0)$ in (8-10) of multiplicity r and one has obtained approximations

$$\beta_j^{(m)}(\epsilon) = \beta_0 + \sum_{k=1}^{m} \alpha_{jk}\epsilon^{k/q} \qquad j = 1, 2, \ldots, r$$

to the functions $\beta_j(\epsilon)$ in (8-12), $\beta_j^{(m)}(\epsilon) - \beta_j(\epsilon) \equiv 0$ (trunc $\epsilon^{m/q}$), $j = 1, 2, \ldots, r$. If each $\beta_j^{(m)}(\epsilon)$ is purely imaginary and $\beta_j^{(m)}(\epsilon) \neq \beta_k^{(m)}(\epsilon)$, for $\epsilon \neq 0$, but sufficiently small, then the exact functions $\beta_j(\epsilon)$ in (8-12) are purely imaginary and distinct for $\epsilon \neq 0$ and sufficiently small. That is, there are r characteristic exponents of system (8-1) which are purely imaginary and distinct for $\epsilon \neq 0$ and sufficiently small.

Proof. Each of the numbers $\beta_j^{(m)}(\epsilon)$ generates an exact solution of (8-11), say $\beta_j(\epsilon)$ and Im $\beta_j(\epsilon) \neq$ Im $\beta_k(\epsilon)$, $j \neq k$, $j, k = 1, 2, \ldots, r$, for $\epsilon \neq 0$ and sufficiently small. But to each $\beta_j(\epsilon)$ there corresponds a characteristic multiplier of (8-1) given by exp $(\lambda_1 + \epsilon\beta_j(\epsilon))T$. But, since (8-1) is reciprocal and real and Im $\beta_j(\epsilon) \neq$ Im $\beta_k(\epsilon)$, $\epsilon \neq 0, j \neq k$, the numbers $\beta_j(\epsilon)$ must be purely imaginary, as was to be shown.

Notice that the proof of the above theorem did not depend upon how the approximations to the exact characteristic exponents were obtained. Therefore, if the original system is reciprocal and by any method of successive approximations one finds that the approximate characteristic exponents are distinct and purely imaginary, then the exact characteristic multipliers are distinct and purely imaginary. This result shows that one can, in general, decide the behavior of the characteristic exponents of reciprocal systems by looking at only a finite number of successive approximations even though the characteristic exponents are purely imaginary. This result has been recently pointed out by Yakubovich [3] for canonical systems (8-1), i.e., systems which satisfy condition (*b*) of Theorem 8-4. For applications of this same idea to systems which satisfy condition (*a*) of Theorem 8-4, see Hale [3].

Now, let us consider the system of equations

$$\begin{aligned} \ddot{y}_j + \sigma_j^2 y_j &= \epsilon \sum_{k=1}^{\mu} (\varphi_{jk} y_k + \psi_{jk} \dot{y}_k) + \epsilon \sum_{k=\mu+1}^{n} \psi_{jk} y_k \qquad j = 1, 2, \ldots, \mu \\ \dot{y}_j &= \epsilon \sum_{k=1}^{\mu} (\varphi_{jk} y_k + \psi_{jk} \dot{y}_k) + \epsilon \sum_{k=\mu+1}^{n} \psi_{jk} y_k \qquad j = \mu + 1, \ldots, n \end{aligned} \tag{8-17}$$

where $\sigma_j > 0$, $j = 1, 2, \ldots, \mu$, and the functions φ_{jk}, ψ_{jk} satisfy the same conditions as the matrix Φ of (8-1). Let $\Phi = (\varphi_{jk})$, $\Psi = (\psi_{jk})$ and suppose that the matrices Φ, Ψ can be partitioned as

$$\begin{aligned} \Phi &= (\Phi_{jk}) \qquad j = 1, 2, 3, \qquad k = 1, 2 \\ \Psi &= (\Psi_{jk}) \qquad j, k = 1, 2, 3 \\ \Phi_{jk}(t) &= (-1)^{k+j}\Phi_{jk}(-t) \qquad \Psi_{jk}(t) = (-1)^{k+j+1}\Psi_{jk}(-t) \end{aligned} \tag{8-18}$$

where Φ_{11}, Ψ_{11} are $p \times p$ matrices Φ_{22}, Ψ_{22} are $(\mu - p) \times (\mu - p)$ matrices, and Ψ_{33} is an $(n - \mu) \times (n - \mu)$ matrix. One can show that this system is reciprocal, but it is not apparent how to discuss in a simple way the $n - \mu$ characteristic multipliers of (8-17) which are equal to 1 for $\epsilon = 0$. By using the method of successive approximations mentioned at the beginning of this section, one can prove the following theorem.

Theorem 8-6. If $\sigma_j \not\equiv 0 \pmod{\omega}$, $j = 1, 2, \ldots, \mu$, and system (8-17) satisfies (8-18), then there are $n - \mu$ characteristic exponents of (8-17) which may be taken equal to zero; moreover, there are $n - \mu$ linearly independent periodic solutions of (8-18) of period T.

The proof of this is very easy if one observes that system (8-17) satisfying (8-18) when transformed to a system of first-order equations of the form

$$\dot{w} = Aw + \epsilon R(t)w$$

$A = \operatorname{diag}(0_{n-\mu}, B)$, has property E (see Definition 6-1) with respect to a matrix Q of the form $Q = \operatorname{diag}(I_\mu, Q_1)$, $Q_1^2 = I$. Since the assumption on σ_j implies that no solution (except the trivial one) of $x' = Bx$ is periodic, the result follows from Corollary 6-1.

Corollary 8-3. If $2\sigma_j \not\equiv 0$, $\sigma_j \pm \sigma_k \not\equiv 0 \pmod{\omega}$, $j \neq k$, $j, k = 1, 2, \ldots$, then all solutions of (8-17) are bounded in $(-\infty, +\infty)$, if ϵ is sufficiently small.

Proof. System (8-17), (8-18) is reciprocal and one merely applies Theorems 8-6 and 8-2.

If some of the conditions of the above corollary are not satisfied, then one can use Theorem 8-5 to discuss the exponents. Also, from Theorem 8-3, we know that, if system (8-1) is reciprocal and all the characteristic multipliers are distinct for $\epsilon = 0$, then all the solutions of (8-1) are bounded in $(-\infty, +\infty)$ for ϵ sufficiently small. However, in the applications, it is very important to know in the (ω,ϵ)-plane the curves of transition from a region of stability to instability. For ϵ very small, these curves can be obtained by the above procedure (see Hale [3]). By a different procedure, Yakubovich [3] has also obtained these transition curves.

The following example illustrates many of the points we have mentioned in this chapter.

Consider the system of two second-order equations

$$\ddot{v} + Av = 2\epsilon(\cos t)\Phi v \tag{8-19}$$

where $v = \operatorname{col}(v_1, v_2)$, $A = \operatorname{diag}[\sigma^2, (\sigma + \epsilon^2\gamma)\sigma]$, $\sigma > 0$, ϵ, γ real parameters,

$$\Phi = \begin{bmatrix} p & 1 \\ q & r \end{bmatrix}$$

and p, q, r are real constants. If $2\sigma \not\equiv 0 \pmod{1}$, we want to determine conditions on σ, p, q, r, γ so that the solutions of (8-19) are bounded in $-\infty < t < \infty$.

Let us make the transformation of variables $v_j = u_{2j-1}$, $\dot{v}_j = u_{2j}$, $j = 1, 2$, and

$$u_{2j-1} = \frac{1}{2i\sigma}(w_j + w_{2+j}) \qquad u_{2j} = \frac{1}{2}(w_j - w_{2+j}) \qquad j = 1, 2$$

in (8-19). If $w = \text{col}\,(w_1,w_2,w_3,w_4)$, then

$$\dot{w} = Cw + \epsilon^2 Dw + \epsilon(\cos t)\Psi w \tag{8-20}$$

where C, Ψ, D are 4×4 matrices

$$\begin{aligned} C &= \text{diag}\,(i\sigma I_2, -i\sigma I_2),\ I_2 = \text{diag}\,(1,1) \\ \Psi &= \frac{1}{i\sigma}\begin{bmatrix} \Phi & \Phi \\ -\Phi & -\Phi \end{bmatrix} \qquad D = \frac{i\gamma}{2}\begin{bmatrix} P & P \\ -P & -P \end{bmatrix} \\ P &= \text{diag}\,(0,1) \end{aligned} \tag{8-21}$$

If we let $w = e^{i(\sigma+\epsilon\beta)t}z$, then

$$\dot{z} = (C - i\sigma I_4)z - i\epsilon\beta z + \epsilon^2 Dz + \epsilon(\cos t)\Psi z$$

or, if $z = \text{col}\,(x,y)$ where x, y are 2 vectors, then

$$\begin{aligned} \dot{x} &= i\epsilon\left(-\beta I_2 + \frac{\epsilon\gamma}{2}P - \frac{\cos t}{\sigma}\Phi\right)x + i\epsilon\left(\frac{\epsilon\gamma}{2}P - \frac{\cos t}{\sigma}\Phi\right)y \\ \dot{y} &= -2i\sigma y - i\epsilon\left(\frac{\epsilon\gamma}{2}P - \frac{\cos t}{\sigma}\Phi\right)x - i\epsilon\left(\beta I_2 + \frac{\epsilon\gamma}{2}P - \frac{\cos t}{\sigma}\Phi\right)y \end{aligned} \tag{8-22}$$

To determine the two characteristic exponents close to $i\sigma$, we must determine β in such a way that (8-22) has a periodic solution. Furthermore, for the solutions of (8-19) to be bounded in $-\infty < t < \infty$, we must have β real (since our original system is a real system, the other two characteristic exponents will also be purely imaginary). The relation which ensures that β be real will be the desired condition on σ, p, q, r, γ.

To obtain a periodic solution of (8-22), we apply the method of successive approximations (7-1), (7-2). If $a = \text{col}\,(a_1,a_2)$ is an arbitrary 2 vector, and $x^{(0)} = a$, $y^{(0)} = 0$, then, from (7-1)

$$\begin{aligned} x^{(1)} &= \left[I_2 - \frac{i\epsilon}{\sigma}(\sin t)\Phi\right]a \\ y^{(1)} &= \frac{\epsilon}{2\sigma}\left(\frac{e^{it}}{2\sigma + 1} + \frac{e^{-it}}{2\sigma - 1}\right)\Phi a \end{aligned}$$

Consequently, from (7-2), the approximate determining equations corresponding to the approximation $z^{(1)} = \text{col}\,(x^{(1)},y^{(1)})$ are

$$\left[-\beta I_2 + \frac{\epsilon\gamma}{2}P - \frac{\epsilon}{\sigma(4\sigma^2 - 1)}\Phi^2\right]a = 0 \tag{8-23}$$

The solutions β of this equation which correspond to nontrivial solutions a are the eigenvalues of the matrix

$$\epsilon\frac{\gamma}{2}P - \frac{\epsilon}{\sigma(4\sigma^2 - 1)}\Phi^2$$
$$= -\frac{\epsilon}{\sigma(4\sigma^2 - 1)}\begin{bmatrix} p^2 + q & p + r \\ q(p + r) & q + r^2 - \sigma\gamma(4\sigma^2 - 1)/2 \end{bmatrix}$$

These eigenvalues will be real if the discriminant

$$[p^2 - r^2 + \sigma\gamma(4\sigma^2 - 1)/2]^2 + 4q(p + r)^2 \tag{8-24}$$

is nonnegative and will be real and distinct if this expression is positive.

Since $\cos t$ is an even function of t, system (8-19) is reciprocal. Therefore, if σ, γ, p, q, r are related in such a way that (8-24) is positive, then from Theorem 8-5 it follows that the solutions are bounded for ϵ sufficiently small. Moreover, if $p + r \neq 0$, $q > 0$, then for any γ, σ the solutions of (8-19) are bounded for ϵ sufficiently small since (8-24) is >0.

On the other hand, if $p + r \neq 0$, $q < 0$, then there always exist two distinct values of γ which make (8-24) equal to zero. Consequently one can always choose γ so that the expression in (8-24) is <0. This implies the solutions β of (8-23) have a nonzero imaginary part, which in turn implies that one of the characteristic exponents close to $i\sigma_1$ has a positive real part and thus there are unbounded solutions of (8-19) for every $\epsilon \neq 0$.

Let us now reinterpret these results. Consider the system

$$\ddot{v} + A^*(\epsilon)v = 2\epsilon(\cos t)\Phi v \tag{8-25}$$

where Φ is the same matrix as in (8-19) and $A^*(\epsilon) = \text{diag}\,[\sigma^2,\sigma_2^2(\epsilon)]$, and $\sigma_2(\epsilon)$ is an analytic function of ϵ at $\epsilon = 0$ and $\sigma_2(0) = \sigma$. If σ is a fixed real number, the point $(\sigma,0)$ in the (σ_2,ϵ)-plane is usually called a *resonance point* of system (8-25). A basic problem is to determine the curves in the (σ_2,ϵ)-plane, if they exist, which separate the regions of stability from the regions of instability of system (8-25). The above analysis on (8-19) shows the following: If in (8-19), $p + r \neq 0$, $q > 0$, then there is no region of instability in a neighborhood of the point $(\sigma,0)$. If $p + r \neq 0$, $q < 0$, then, in general, there is a region of instability in a neighborhood and the curves which separate the region of instability from the region of stability are given by

$$\sigma_2 = \sigma + \epsilon^2\gamma_j + 0(\epsilon^3)$$

where γ_j, $j = 1, 2$, are values of γ which make (8-24) equal to zero (see Fig. 8-1).

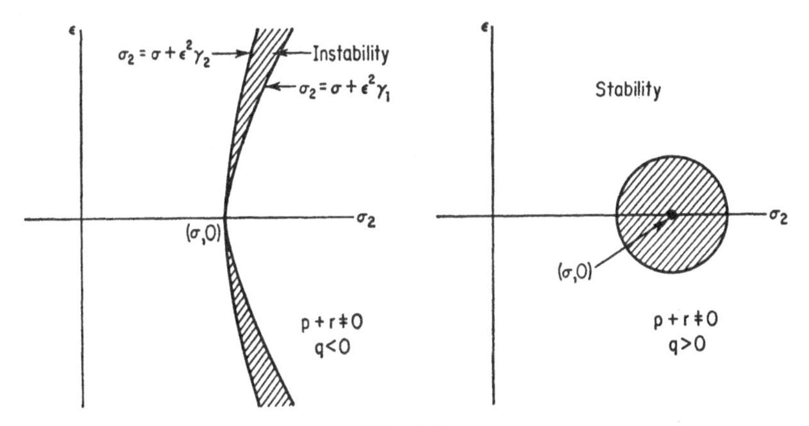

Fig. 8-1

One final remark about this example. Notice that in (8-19) we took $\sigma_2^2 = (\sigma + \epsilon^2\gamma)\sigma$. Why did we not take $\sigma_2^2 = (\sigma + \epsilon\gamma)\sigma$? One could do this and obtain the first approximation to the determining equations of the form $[-\beta I_2 + (\gamma/2)P]a = 0$, and the matrix $(\gamma/2)P$ always has real eigenvalues for all real γ. But, for $\gamma = 0$, both eigenvalues are zero and Theorem 8-5 would not apply for $\gamma = 0$. Consequently, no information is obtained about the transition curves from stability to instability (if they exist) since the solutions of (8-25) are always bounded along all straight lines passing through $(\sigma,0)$ except the vertical line if ϵ is sufficiently small. Finally, one can only assert that if there is an instability region in a neighborhood of $(\sigma,0)$ the transition curves must be tangent at $(\sigma,0)$ to a higher order in ϵ.

In the discussion of this section concerning the characteristic exponents, a basic assumption has been that each eigenvalue of the matrix B_1 in (8-1) has simple elementary divisors. If this is not the case, then one can still reduce the problem of finding characteristic exponents to a problem of finding periodic solutions of an equation in the standard form (6-1). This reduction generalizes the procedure used in Sec. 7-8 and then the characteristic exponents are determined as a function of some fractional power of the small parameter ϵ. A detailed discussion of linear periodic systems of the form (8-1) with the eigenvalues of the matrix B_1 having nonsimple elementary divisors can be found in Imaz [1].

We now reproduce some of the results of Imaz. Consider the very special case of system (8-1),

$$\dot{w} = Cw + \epsilon\Phi(t)w \tag{8-26}$$

where

$$C = \begin{bmatrix} 0 & 1 & \cdots & 0 \\ 0 & 0 & \cdots & 0 \\ \cdot & \cdot & & \cdot \\ \cdot & \cdot & & \cdot \\ \cdot & \cdot & & 1 \\ 0 & 0 & \cdots & 0 \end{bmatrix} \qquad \text{and} \qquad w = \begin{bmatrix} w_1 \\ w_2 \\ \cdot \\ \cdot \\ \cdot \\ w_p \end{bmatrix}$$

and suppose $\epsilon > 0$ and the mean value of the matrix $\Phi(t)$ is zero, that is,

$$\frac{1}{T}\int_0^T \Phi(t)\,dt = 0 \tag{8-27}$$

If $\lambda = \epsilon^{1/p}$, and

$$P(\lambda) = \operatorname{diag}\,(1, \lambda, \lambda^2, \ldots, \lambda^{p-1}) \qquad \lambda > 0 \tag{8-28}$$

then the transformation $w = P(\lambda)z$ yields the equivalent system

$$\dot{z} = \lambda\Psi(t,\lambda)z \tag{8-29}$$

where

$$\Psi(t,\lambda) = \begin{bmatrix} \lambda^{p-1}\varphi_{11} & \lambda^{p}\varphi_{12} & \cdots & \lambda^{2p-2}\varphi_{1p} \\ \lambda^{p-2}\varphi_{21} & \lambda^{p-1}\varphi_{22} & \cdots & \lambda^{2p-3}\varphi_{2p} \\ \cdot & & & \\ \cdot & & & \\ \cdot & & & \\ \varphi_{n1} & \lambda\varphi_{n2} & \cdots & \lambda^{p-1}\varphi_{pp} \end{bmatrix} \tag{8-30}$$

and φ_{jk} is the (j,k)th element of the matrix Φ. System (8-29) is now a special case of system (8-5) (with ϵ replaced by λ) and the preceding theory applies.

By applying the method of successive approximations (7-1), (7-2), one finds that the matrix $G(\beta,\lambda)$ in (8-9) up through terms of order λ^p

is given by

$$G(\beta,\lambda) = \begin{bmatrix} 0 & 1 & 0 & \cdots & 0 \\ 0 & 0 & 1 & \cdots & 0 \\ \cdot & & & & \\ \cdot & & & & \\ \cdot & & & & \\ 0 & 0 & 0 & \cdots & 1 \\ \lambda^p M & 0 & 0 & \cdots & 0 \end{bmatrix} \tag{8-31}$$

$$M = \frac{1}{T}\int_0^T \Big\{\varphi_{p1}\Big[\int \varphi_{11} + \iint \varphi_{21} + \cdots + \int \cdots \int \varphi_{p1}\Big] + \varphi_{p2}\Big[\int \varphi_{21} + \iint \varphi_{31} + \cdots + \int \cdots \int \varphi_{p1}\Big] + \cdots + \varphi_{pp}\int \varphi_{p1}\Big\}\, dt$$

where each of the signs $\int$ denotes taking the unique primitive of mean value zero. Consequently, the characteristic exponents β of (8-26) are given by the solutions of the equation

$$\beta^p - \lambda^p M + 0(\lambda^{p+1}) = 0 \tag{8-32}$$

By using the Newton polygon (see Imaz [1]), one can now prove the following theorem.

Theorem 8-7. Suppose system (8-26) satisfies (8-27), M is defined by (8-31), and $\epsilon > 0$. The following conclusions hold:

1. If either $M \neq 0$, $p > 2$ or $M > 0$, $p = 2$, then, for any ϵ sufficiently small, there are unbounded solutions of (8-26).
2. If $M < 0$, $p = 2$, and system (8-26) is reciprocal, then, for $\epsilon \neq 0$ sufficiently small, the solutions of (8-26) are bounded in $(-\infty, \infty)$.

We do not wish to discuss the details of the Newton polygon, but the validity of Theorem 8-7 is due to the fact that the "first approximation" to the solutions of (8-32) is given by

$$\beta^p = \lambda^p M$$

or $\beta = \lambda \sqrt[p]{M}$ where $\sqrt[p]{M}$ is any one of the pth roots of M. If $p > 2$ these cannot all lie in the left half plane and if $M > 0$, $p = 2$, then one is positive and one negative. If $M < 0$, $p = 2$, then both roots are distinct and lie on the imaginary axis and the reciprocal property implies boundedness (Theorem 8-5). Notice the constant M is almost always $\neq 0$.

Many other results and examples may be found in the paper of Imaz [1].

9: Periodic Solutions of Nonautonomous Systems

The purpose of this chapter is to show how a general class of nonautonomous systems may be reduced to the standard form (6-1), (6-2). The special form of the determining equations (6-11), or (6-14), for this case will also be given. Further, a class of equations which has property E (see Definition 6-1) will be exhibited. Consider the real system of nonautonomous differential equations

$$\dot{w} = Cw + \epsilon f(t,w,\epsilon) \tag{9-1}$$

where ϵ is a real parameter, w is an n vector, C is an $n \times n$ constant matrix, $f(t,w,\epsilon)$ is continuous in t and has continuous first and second partial derivatives with respect to w, ϵ in $-\infty < t < +\infty$, $0 \leqq \|w\| \leqq R$, $0 \leqq |\epsilon| \leqq \epsilon_0$, $\epsilon_0 > 0$, periodic in t of period T_1. Suppose that $C = \operatorname{diag}(B_1,B_2)$ where B_1 is a $p \times p$ matrix such that every solution of the equation $\dot{x} = B_1x$ is periodic with period L, where L/T_1 is rational, and B_2 is an $(n - p) \times (n - p)$ matrix such that no solution of the equation $\dot{x} = B_2x$ is periodic of period T, where T is the least common multiple of L and T_1.

By an appropriate change of coordinates, the matrix C can always be assumed to have this form if the following condition is satisfied: If $T_1 = 2\pi/\omega$, $\omega > 0$, then the eigenvalues of C which are in rational ratio with ω have simple elementary divisions.

Let $w = \operatorname{col}(u,v)$, $f = \operatorname{col}(g,h)$ where u, g are p vectors and consider the transformation of variables

$$u = e^{B_1t}x \qquad v = y \tag{9-2}$$

which is periodic in t of period L. The new equations for x, y are

$$\begin{aligned}\dot{x} &= \epsilon e^{-B_1 t} g(t, e^{B_1 t}x,\ y,\ \epsilon) \stackrel{\text{def}}{=} \epsilon X(t,x,y,\epsilon) \\ \dot{y} &= B_2 y + \epsilon h(t, e^{B_1 t}x,\ y,\ \epsilon) \stackrel{\text{def}}{=} B_2 y + \epsilon Y(t,x,y,\epsilon)\end{aligned} \tag{9-3}$$

where X, Y are periodic in t of period T, the least common multiple of (L,T_1), and satisfy all the conditions of system (6-1). System (9-3) is a special case of system (6-1), (6-2) with $z = \text{col}\,(x,y)$, $Z = \text{col}\,(X,Y)$.

Consequently, since (9-2) is a periodic transformation of period T, the theory developed in Chap. 6 can be used to obtain a periodic solution of (9-1). To see that we obtain the same results that have been obtained by other methods, let us write the determining equations in terms of the original vector f and the matrix C in (9-1).

Let $b < R$ be a given constant and let a be a given real constant p vector, $\|a\| \leqq b$. For this a, Theorem 6-1 implies the existence of a periodic $z(t,a,\epsilon)$, $z(t,a,0) = \text{col}\,(a,0)$ $z = \text{col}\,(x,y)$, associated with system (9-3). Furthermore, this function is obtained by the method of successive approximations (6-9). The determining equations (6-11) are given by (6-14) which in terms of f, C in (9-3) are

$$F(a,\epsilon) \stackrel{\text{def}}{=} \int_0^T e^{-B_1 t} g(t, e^{B_1 t}x(t,a,T,\epsilon),\ y(t,a,T,\epsilon),\ \epsilon) = 0 \tag{9-4}$$

Consequently, if there exists a function $a(\epsilon)$, $\|a(\epsilon)\| \leqq b < R$, $0 \leqq |\epsilon| \leqq \epsilon_2$ such that $F(a(\epsilon),\ \epsilon) = 0$, then there is a periodic solution of (9-1) of period T given by $w = \text{col}\,(u,v)$, $u = [\exp\,(B_1 t)]x(t,\ a(\epsilon),\ T,\ \epsilon)$, $v = y(t,\ a(\epsilon),\ T,\ \epsilon)$. Notice that $F(a,0)$ can be calculated without knowing anything except a and, in fact, is given explicitly by

$$F(a,0) = \int_0^T e^{-B_1 t} g(t, e^{B_1 t}a, 0, 0)\, dt \tag{9-5}$$

From the implicit-function theorem, it follows that, if there exists a real constant p vector a_0, $\|a_0\| < R$, such that

$$F(a_0,0) = 0 \qquad \det\left[\frac{\partial F\,(a_0,0)}{\partial a}\right] \neq 0 \tag{9-6}$$

then there exists an $\epsilon_2 > 0$, such that there is a periodic solution $w(t,a(\epsilon),\epsilon)$, $a(0) = a_0$, $w = \text{col}\,(u,v)$, of (9-1) of period T and $u(t,a_0,0) = [\exp\,(B_1 t)]a_0$, $v(t,a_0,0) = 0$.

This result is precisely the same as is obtained by taking the first approximation in Poincaré's perturbation procedure (see Coddington and Levinson [1, p. 356]). If the determinant in (9-6) is equal to zero, then by the method of successive approximations given in (6-9), one can obtain higher-order terms of $F(a,\epsilon)$ in (9-4) and thus other conditions for periodic solutions.

Also, as was seen in some of the examples in Chap. 7, the property E in Definition 6-1 can be very useful to show that, for certain types of p vectors a (see Theorem 6-6), some of the equations in (9-4) are equal to zero. Consequently, we wish to exhibit a class of equations (9-1) which have property E with respect to a simple matrix Q.

Suppose that all the eigenvalues of the matrix C in (9-1) have simple elementary divisors and are either purely imaginary or zero. Then, without loss of generality, we can assume that

$$\begin{gathered} C = \operatorname{diag}(0_\mu, C_1, \ldots, C_k) \\ 0_\mu \text{ is the } \mu \times \mu \text{ zero matrix} \\ C_j = \begin{bmatrix} 0 & 1 \\ -\sigma_j^2 & 0 \end{bmatrix} \qquad \sigma_j > 0 \qquad j = 1, 2, \ldots, k \end{gathered} \tag{9-7}$$

Lemma 9-1. Suppose C in system (9-1) satisfies condition (9-7) and let $\alpha_1, \ldots, \alpha_k$ be constants which are either $+1$ or -1 (not all the α_j need be equal). If I_μ is the $\mu \times \mu$ identity matrix

$$Q = \operatorname{diag}(I_\mu, \alpha_1 P, \ldots, \alpha_k P) \qquad P = \operatorname{diag}(1, -1) \tag{9-8}$$

and

$$Qf(-t, Qw, \epsilon) = -f(t, w, \epsilon) \tag{9-9}$$

then system (9-1) and also system (9-3) have property E with respect to Q.

Proof. It is clear that $Q^2 = I$, $QP_0 = P_0Q$. Furthermore, $QCQw = -Cw$ for all z, if and only if $CQ = -QC$, which is true if and only if $C_j\alpha_jP = -\alpha_jPC_j$, $j = 1, 2, \ldots, k$. This last relation is easily checked by calculation and thus system (9-1) has property E with respect to Q. Since $C_jP = -PC_j$, $j = 1, 2, \ldots, k$, it is also easily verified that system (9-3) has property E with respect to Q.

If the conditions of this lemma are satisfied, then it follows from Theorem 6-6, that the first μ components of the vector $F(a,\epsilon)$ in (9-4) are identically zero in the vector a. Also, if C is given by (9-7) and B_2 in (9-1) is given by $B_2 = \operatorname{diag}(C_1, \ldots, C_k)$, then there is a μ-parameter family of periodic solutions of (9-1).

If $B_2 = \operatorname{diag}(C_{\nu+1}, \ldots, C_k)$, $\nu > 0$, then for any $(\mu + 2\nu)$ vector a for which $Qa^* = a^*$, $a^* = (a,0)$, at most ν of the components of $F(a,\epsilon)$ in (9-4) are nonzero, and the possible nonzero ones are easily obtained from the sign of the α_j, $j = 1, 2, \ldots, \nu$.

For systems of second-order equations

$$\ddot{x}_j + \sigma_j^2 x_j = \epsilon f_j(t, x_1, \dot{x}_1, \ldots, x_n, \dot{x}_n) \qquad j = 1, 2, \ldots, n \quad (9\text{-}10)$$

the corresponding system of order $2n$ obtained by letting $x_j = z_{2j-1}$, $\dot{x}_j = z_{2j}$, $j = 1, 2, \ldots, n$, will have property E with respect to Q in (9-8) if

$$f_j(-t, \alpha_1 x_1, -\alpha_1 \dot{x}_1, \ldots, \alpha_n x_n, -\alpha_n \dot{x}_n) = \alpha_j f_j(t, x_1, \dot{x}_1, \ldots, x_n, \dot{x}_n)$$
$$j = 1, 2, \ldots, n \quad (9\text{-}11)$$

Notice how easy it is to check this condition for Secs. 7-2 and 7-3. There, one can take either $\alpha_1 = +1$, or $\alpha_1 = -1$.

We do not exploit this situation any further but refer to the papers of Cesari and Hale [2], Hale [12,6], Gambill and Hale [1] for applications and examples making use of the above results.

In case the eigenvalues of the matrix C in (9-1) which are in rational ratio with ω, $T_1 = 2\pi/\omega$, do not have simple elementary divisors, one can always reduce the equation to "standard" form by a procedure which generalizes that used in Sec. 7-8 (see the last part of Chap. 8).

For example, suppose that C in (9-1) is given by $C = \text{diag}\,(B_1,B_2)$ where B_2 satisfies the same conditions as before and

$$B_1 = \begin{bmatrix} 0 & 1 & & & & & \\ & 0 & 1 & & & & \\ & & 0 & \cdot & & & \\ & & & \cdot & \cdot & & \\ & & & & \cdot & \cdot & \\ & & & & & \cdot & \cdot \\ & & & & & 0 & 1 \\ & & & & & & 0 \end{bmatrix}$$

that is, B_1 has all elements zero except the elements on the superdiagonal and those on the superdiagonal are equal to 1. If $\epsilon > 0$, let $\lambda = \epsilon^{1/p}$,

$$P(\lambda) = \text{diag}\,(1, \lambda, \lambda^2, \ldots, \lambda^{p-1})$$

and the transformation $z_1 = P(\lambda)y_1$, $z_2 = y_2$ leads to a system of equations in y_1, y_2 which are in standard form (6-1), (6-2).

As a final remark, consider systems of the form

$$\dot{u} = Cu + \epsilon f(t,u,\epsilon) + g(t) \qquad (9\text{-}12)$$

where C, f satisfy the same conditions as in system (9-1), g is periodic in t of period T, and the linear nonhomogeneous system

$$\dot{u} = Cu + g(t) \qquad (9\text{-}13)$$

has a periodic solution. Systems of this type occur more frequently in the applications than system (9-1). However, if we let $u_0(t)$ be any periodic solution of (9-13), then the transformation $u = u_0(t) + w$ leads to a system of equations of the form (9-1).

The system (9-13) has a periodic solution if and only if

$$\int_0^T [\exp(-B_1 t)] g^{(1)}(t)\, dt = 0 \tag{9-14}$$

where $g = \text{col}\,(g^{(1)},g^{(2)})$, $g^{(1)}$ a p vector. In fact, the transformation (9-2) applied to (9-13) yields

$$\begin{aligned} \dot{x} &= [\exp(-B_1 t)] g^{(1)}(t) \\ \dot{y} &= B_2 y + g^{(2)}(t) \end{aligned}$$

and the result follows immediately.

The discussion above has shown how to obtain periodic solutions of a nonlinear differential equation. We now indicate how to decide whether or not such a periodic solution is stable. The analysis depends upon calculating the first approximation to the characteristic exponents and then using Theorem 4-1.

Consider the system of second-order equations

$$\begin{aligned} \ddot{w} + Aw &= \epsilon f(w,\dot{w},t) \\ A &= \text{diag}\,(\sigma_1^2, \ldots, \sigma_n^2) \end{aligned} \tag{9-15}$$

where each $\sigma_j > 0$ and it is assumed that $f(w,\dot{w},t)$ contains t explicitly, is periodic in t of period $T = 2\pi/\omega$, and is continuous in w, $\dot{w}$, t and has continuous first and second partial derivatives with respect to w, $\dot{w}$, for all $t \geqq 0$ and w, $\dot{w}$ in some region.

Furthermore we suppose that

$$\sigma_1 = \omega \qquad \sigma_j \neq m\omega \qquad \sigma_j - \sigma_k \neq m\omega \qquad j, k = 2, 3, \ldots, n \qquad m = 0, \pm 1, \pm 2, \ldots \tag{9-16}$$

and that there exists a periodic solution $w^*(t,\epsilon)$ of (9-15) of period T, continuous in t, ϵ with a continuous first derivative with respect to ϵ, such that

$$w^*(t,0) = \text{col}\,(a \sin(\omega t + \varphi), 0, \ldots, 0) \qquad a, \varphi \text{ constant} \tag{9-17}$$

Define the number S by the following relation:

$$\begin{aligned} (2T)^2 S = {} & \left(\int_0^T f_{1\dot{w}_1}\, dt\right)^2 + \frac{1}{\omega^2}\left(\int_0^T f_{1w_1}\, dt\right)^2 \\ & - \left(\int_0^T \left[\frac{1}{\omega} f_{1w_1} \cos 2\omega t - f_{1\dot{w}_1} \sin 2\omega t\right]\right)^2 \\ & - \left(\int_0^T \left[\frac{1}{\omega} f_{1w_1} \sin 2\omega t + f_{1\dot{w}_1} \cos 2\omega t\right] dt\right)^2 \end{aligned} \tag{9-18}$$

where f_{1w_1}, $f_{1\dot{w}_1}$ represent the first partial derivatives of the first coordinate of f with respect to w_1, $\dot{w}_1$ respectively evaluated at $w^*(t,0)$, where $w^*(t,0)$ is given in (9-17).

Theorem 9-1. If system (9-15) satisfies (9-16) and has a periodic solution $w(t,\epsilon)$ which satisfies (9-17), then for ϵ sufficiently small, this periodic solution is asymptotically stable if the following conditions are satisfied:

$$\epsilon \int_0^T f_{j\dot{w}_j}[w^*(t,0), \dot{w}^*(t,0),t]\, dt < 0 \qquad j = 1, 2, \ldots, n \qquad (a)$$

$$S > 0 \qquad \text{where } S \text{ is defined in (9-18)} \qquad (b)$$

Proof. The linear variational equations associated with the solution $w^*(t,\epsilon)$ are given by

$$\begin{aligned} \ddot{y} + Ay &= \epsilon\Phi(t,\epsilon)y + \epsilon\Psi(t,\epsilon)\dot{y} \\ \Phi(t,\epsilon) &= \frac{\partial f\,[w^*,\dot{w}^*,t]}{\partial w} \qquad \Psi(t,\epsilon) = \frac{\partial f\,[w^*,\dot{w}^*,t]}{\partial \dot{w}} \end{aligned} \qquad (9\text{-}19)$$

From Theorem 4-1, to prove the theorem, it is sufficient to show that the characteristic exponents of (9-19) have negative real parts.

If $y = \text{col}\,(y_1, \ldots, y_n)$, the transformation $y_j = x_{2j-1}$, $\dot{y}_j = x_{2j}$, $j = 1, 2, \ldots, n$, yields the equivalent system of equations

$$\begin{aligned} \dot{x} &= Bx + \epsilon F(t,\epsilon)x \\ x &= \text{col}\,(x_1, \ldots, x_{2n}) \\ B &= \text{diag}\,(B_1, \ldots, B_n) \\ B_j &= \begin{bmatrix} 0 & 1 \\ -\sigma_j^2 & 0 \end{bmatrix} \qquad j = 1, 2, \ldots, n \end{aligned} \qquad (9\text{-}20)$$

and $F(t,\epsilon)$ is a $2n \times 2n$ matrix depending upon $\Phi(t,\epsilon)$, $\Psi(t,\epsilon)$.

For $\epsilon = 0$, the characteristic multipliers of (9-20) are $\exp \pm i\sigma_j T$, $j = 1, 2, \ldots, n$. Hypothesis (9-16) implies that these multipliers are all distinct except the two which correspond to σ_1, since $\sigma_1 = \omega$, $T = 2\pi/\omega$, and these two are equal to 1. Furthermore, to obtain these multipliers, for $\epsilon \neq 0$ we can apply the procedure of Chap. 8. We want to obtain the first approximation to the corresponding characteristic exponents and then apply Corollary 8-1. If we make a transformation of variables to transform the matrix B_j to a diagonal form then it is clear that the matrix $G(\beta,0)$ in formula (8-10) will depend only upon the (j,j)th elements $\varphi_{jj}(t,0)$, $\psi_{jj}(t,0)$ of the matrices $\Phi(t,\epsilon)$, $\psi(t,\epsilon)$. Consequently, to determine sufficient conditions for the characteristic exponents of (9-19) to have negative real parts, we need only consider the second-order systems

$$\dot{z} = B_j z + \epsilon C_j(t) z$$
$$B_j = \begin{bmatrix} 0 & 1 \\ -\sigma_j^2 & 0 \end{bmatrix} \quad C_j = \begin{bmatrix} 0 & 0 \\ \varphi_{jj}(t,0) & \psi_{jj}(t,0) \end{bmatrix} \quad j = 1, 2, \ldots, n \tag{9-21}$$

If $z = \text{col}\,(z_1,z_2)$, $z_1 = (2i\sigma_j)^{-1}(u_1 + u_2)$, $z_2 = 2^{-1}(u_1 - u_2)$, then

$$\begin{aligned} \dot{u}_1 &= i\sigma_j u_1 + \frac{\epsilon}{2}\left[\frac{\varphi_{jj}}{i\sigma_j} + \psi_{jj}\right] u_1 + \frac{\epsilon}{2}\left[\frac{\varphi_{jj}}{i\sigma_j} - \psi_{jj}\right] u_2 \\ \dot{u}_2 &= -i\sigma_j u_2 - \frac{\epsilon}{2}\left[\frac{\varphi_{jj}}{i\sigma_j} + \psi_{jj}\right] u_1 - \frac{\epsilon}{2}\left[\frac{\varphi_{jj}}{i\sigma_j} - \psi_{jj}\right] u_2 \end{aligned} \tag{9-22}$$

For $j \geqq 2$, we have a special case of system (8-1) with $p = 1$, $n = 2$. Consequently, the first approximation to the characteristic exponent $\tau_{2j-1}(\epsilon)$, $\tau_{2j-1}(0) = i\sigma_j$ is

$$\tau_{2j-1}(\epsilon) = i\sigma_j - \frac{i\epsilon}{2\sigma_j}\left[\frac{1}{T}\int_0^T \varphi_{jj}(t,0)\,dt\right] + \frac{\epsilon}{2T}\int_0^T \psi_{jj}(t,0)\,dt$$

and thus

$$\text{Re}\,(\tau_{2j-1}(\epsilon)) = \frac{\epsilon}{2T}\int_0^T \psi_{jj}(t,0)\,dt \qquad j = 2, 3, \ldots, n$$

In the same way one shows that $\text{Re}\,(\tau_{2j}(\epsilon)) = \text{Re}\,(\tau_{2j-1}(\epsilon))$, $\tau_{2j}(0) = -i\sigma_j$, $j = 2, 3, \ldots, n$. From the definition of $\psi_{jj}(t,0)$ in (9-19) and condition (*a*) of Theorem 9-1, it follows that all these characteristic exponents have negative real parts for ϵ sufficiently small.

For $j = 1$, the two characteristic multipliers $\mu_1(\epsilon)$, $\mu_2(\epsilon)$ are equal to 1 for $\epsilon = 0$. System (9-22) is then a special case of (8-1) with $p = 2$, $n = 2$. One then applies Corollary 8-1 and conditions (*a*), (*b*) of Theorem 9-1 imply that the eigenvalues β of the matrix $G(\beta,0)$ in (8-10) have negative real parts. The theorem is proved.

As an exercise, for the system

$$\begin{aligned} \ddot{x}_1 + x_1 &= \epsilon(1 - x_1^2 - x_2^2)\dot{x}_1 + \epsilon p \cos t \\ \ddot{x}_2 + \sigma^2 x_2 &= \epsilon(1 - x_1^2 - x_2^2)\dot{x}_2 \end{aligned}$$

where σ is not equal to an integer, find conditions on p which will ensure that this equation has an asymptotically stable periodic solution which for $\epsilon = 0$ has $x_1 = a \sin t$, $x_2 = 0$.

10: Periodic Solutions of Autonomous Systems

The purpose of the present chapter is to show how the problem of finding periodic solutions of a general class of autonomous differential systems may be reduced to the problem of determining periodic solutions of a nonautonomous system in the standard form (6-1). The special form of the determining equations will also be given. The relationship between property E (see Definition 6-1) and the existence of families of solutions will also be pointed out.

Consider the real system of autonomous differential equations

$$\dot{w} = Cw + \epsilon f(w,\epsilon) \tag{10-1}$$

where ϵ is a real parameter, w is an n vector, C is an $n \times n$ constant matrix, $f(w,\epsilon)$ is continuous and has continuous first and second partial derivatives with respect to w, ϵ for $\|w\| \leqq R$, $0 \leqq |\epsilon| \leqq \epsilon_0$, $R > 0$, $\epsilon_0 > 0$. Suppose that $C = \operatorname{diag}(B_1,B_2)$, where B_1 is a $p \times p$ matrix such that every solution of the equation $\dot{w} = B_1 w$ is periodic of a common period T_1 and B_2 is an $(n - p) \times (n - p)$ matrix such that no solution except $w = 0$ of the equation $\dot{w} + B_2 w$ is periodic of period T_1. By an appropriate change of coordinates, the matrix C can always be assumed in this form if the eigenvalues of the matrix C which correspond to periodic solutions of period T_1 have simple elementary divisors.

The above assumption on B_1 implies that all the eigenvalues of B_1 have simple elementary divisors, are purely imaginary, and are rational multiples of some positive number σ. Without loss in generality, we may also assume that

$$B_1 = \text{diag}\,(0_\mu, C_1, \ldots, C_\nu)$$
$$0_\mu \text{ is the } \mu \times \mu \text{ zero matrix} \tag{10-2}$$
$$C_j = \begin{bmatrix} 0 & 1 \\ -r_j^2\sigma^2 & 0 \end{bmatrix} \qquad j = 1, 2, \ldots, \nu$$

where each r_j is rational, $j = 1, 2, \ldots, \nu$. Let us write B_1 as $B_1(\sigma^2)$ and T_1 as $T_1(\sigma)$. If $\tau^2 = \sigma^2 + \epsilon\beta$, where β is to be determined as a function of ϵ, let $w = \text{col}\,(u,v)$, $f = \text{col}\,(g,h)$ where u, g are p vectors and make the transformation of variables

$$u = e^{B_1(\tau^2)t}x \qquad v = y \tag{10-3}$$

The new equations for x, y are

$$\begin{aligned} \dot{x} &= \epsilon e^{-B_1(\tau^2)t}g(e^{B_1(\tau^2)t}x,y,\epsilon) - \epsilon e^{-B_1(\tau^2)t}B^*(\beta)e^{B_1(\tau^2)t}x \\ \dot{y} &= B_2 y + \epsilon h(e^{B_1(\tau^2)t}x,y,\epsilon) \end{aligned} \tag{10-4}$$

where we have put $B_1(\sigma^2) - B_1(\sigma^2 + \epsilon\beta) = -\epsilon B^*(\beta)$. This system can be written in more compact form as

$$\dot{z} = Az + \epsilon Z(t,z,\epsilon,\beta) \tag{10-5}$$

where $z = \text{col}\,(x,y)$, $A = \text{diag}\,(0,B_2)$, and Z is periodic in t of period $T^* \overset{\text{def}}{=} T_1[(\sigma^2 + \epsilon\beta)^{1/2}]$. Clearly, (10-5) is a special case of system (6-1), (6-2).

We can apply the theory of Chap. 6 to the determination of periodic solutions of (10-5) of period T^*, and since the transformation (10-3) is a periodic transformation of period T^*, this will yield periodic solutions of (10-1). In fact, let $b < R$ be a given constant and let a be a given real constant p vector, $\|a\| \leqq b$. If β is given and $\epsilon_0 > 0$ is so small that $[\exp\,(-B_2T^*) - I]$ is nonsingular, $0 \leqq |\epsilon| \leqq \epsilon_0$, then for this a, β Theorem 6-1 implies the existence of a periodic function $z(t,a,\beta,\epsilon)$, $0 \leqq |\epsilon| \leqq \min\,(\epsilon_1,\epsilon_0)$, $z(t,a,\beta,0) = \text{col}\,(a,0)$ associated with system (10-5). Furthermore, this function is obtained by the method of successive approximations (6-9). The determining equations (6-11) are given in terms of (6-14) which in terms of the vector $f = \text{col}\,(g,h)$, $z = \text{col}\,(x,y)$, and the matrix $C = \text{diag}\,(B_1,B_2)$ in (10-1) are

$$F(a,\beta,\epsilon) \overset{\text{def}}{=} \int_0^{T^*} e^{-B_1(\tau^2)t}[g(e^{B_1(\tau^2)t}x(t,a,\beta,\epsilon), y(t,a,\beta,\epsilon), \epsilon) - B^*(\beta)e^{B_1(\tau^2)t}x]\,dt \tag{10-6}$$

$\tau^2 = \sigma^2 + \epsilon\beta$, $T^* = T_1(\tau)$. Consequently, if there exist functions $\beta(\epsilon)$, $a(\epsilon)$, $\|a(\epsilon)\| \leqq b < R$, $0 \leqq |\epsilon| \leqq \epsilon_2$ such that $F(a(\epsilon),\beta(\epsilon),\epsilon) = 0$ then there is a periodic solution of (10-1) of period $T_1[(\sigma^2 + \epsilon\beta(\epsilon))^{1/2}]$ given by $w = \text{col}\,(u,v)$, where

$$u = e^{B_1(\sigma^2+\epsilon\beta(\epsilon))t}x(t,a(\epsilon),\epsilon),\ v = y(t,a(\epsilon),\beta(\epsilon),\epsilon)$$

and defined for $0 \leqq |\epsilon| \leqq \epsilon_3$, $\epsilon_3 > 0$, sufficiently small. Notice that $F(a,\beta,0)$ may be calculated without knowing anything except the vector a and, in fact, is given explicitly by

$$F(a,\beta,0) = \int_0^{T_1(\sigma)} e^{-B_1(\sigma^2)t}[g(e^{B_1(\sigma^2)t}a,0,0) - B^*(\beta)e^{B_1(\sigma^2)t}a]\,dt \tag{10-7}$$

where

$$\begin{aligned} B^*(\beta) &= \text{diag}\,(0_\mu, D_1, \ldots, D_\nu) \\ &0_\mu \text{ is the } \mu \times \mu \text{ zero matrix} \\ D_j &= \begin{bmatrix} 0 & 0 \\ -r_j^2\beta & 0 \end{bmatrix} \qquad j = 1, 2, \ldots, \nu \end{aligned} \tag{10-8}$$

System (10-6) consists of p equations for $(p + 1)$ unknowns a, β. Hereafter, when we speak of determining these unknowns, we always understand that $p - 1$ of the components of a together with β are determined as functions of the other component of a and ϵ. It is quite natural to take one component of a as arbitrary since the original equation (10-1) is autonomous.

From the implicit-function theorem, it follows that, if there exists a real constant p vector a_0, $\|a_0\| < R$ and a real β_0 such that

$$F(a_0,\beta_0,0) = 0 \qquad \text{rank}\,[\partial F\,(a_0,\beta_0,0)/\partial(a,\beta)] = p \tag{10-9}$$

then there exists an $\epsilon_2 > 0$ such that there is a periodic solution $w(t,a_0,\beta_0,\epsilon)$, $0 \leqq |\epsilon| \leqq \epsilon_2$, of (10-1) of period $T_1[(\sigma^2 + \epsilon\beta(\epsilon))^{1/2}]$, and $\beta(0) = \beta_0$, and $w = \text{col}\,(u,v)$, where $u(t,a_0,\beta_0,0) = \{\exp\,[B_1(\sigma^2)t)]\}a_0$, $v(t,a_0,\beta_0,0) = 0$.

Now suppose that the matrix C in (10-1) satisfies relation (9-7) and the matrix Q is defined by (9-8). Then Lemma 9-1 implies that system (10-1) and system (10-4) have property E with respect to Q and

$$Qf(Qw,\epsilon) = -f(w,\epsilon) \tag{10-10}$$

A simple system of n second-order equations for which the corresponding $2n$-dimensional system of first-order equations has property E with respect to Q is

$$\ddot{x}_j + \sigma_j^2 x_j = \epsilon f_j(x_1, \dot{x}_1, \ldots, x_n, \dot{x}_n, \epsilon) \qquad j = 1, 2, \ldots, n \tag{10-11}$$

with

$$f_j(\alpha_1 x_1, -\alpha_1\dot{x}_1, \ldots, \alpha_n x_n, -\alpha_n\dot{x}_n, \epsilon) = \alpha_j f_j(x_1, \dot{x}_1, \ldots, x_n, \dot{x}_n, \epsilon) \qquad j = 1, 2, \ldots, n \tag{10-12}$$

where each α_j is either $+1$ or -1.

If system (10-1) satisfies relation (10-10), then Theorem 6-6 states that the first μ components of the vector $F(a,\beta,\epsilon)$ will be zero for those a for which $Qa^* = a^*$, $a^* = \text{col}\,(a,0)$. This condition on a forces ν of the components of a to be zero. Also, from Theorem 6-5, at most ν of the remaining components of $F(a,\beta,\epsilon)$ are different from zero. Consequently, the equations $F(a,\beta,\epsilon)$ represent ν equations for the $\mu + \nu + 1$ unknowns consisting of the nonzero components of a and β. Therefore, one in general expects a $(\mu + 1)$-parameter family of periodic solutions of (10-1) if the above symmetry properties are satisfied. Finally, since one can also obtain other periodic solutions by a phase shift, there will generally be a $(\mu + 2)$-parameter family of periodic solutions (see Hale [2,6]).

In fact, it is shown in Hale [6] that if $\sigma_j \not\equiv 0 \pmod{\sigma_l}$, $j \neq l$, $j, l = 1, \ldots, k$, and (10-10) is satisfied, there are k distinct $(\mu + 2)$-parameter families of periodic solutions of (10-1). This result implies that all solutions in a neighborhood of $y = \dot{y} = \ddot{y} = 0$ of the equation

$$\begin{aligned} \dddot{y} + \sigma^2\dot{y} &= \epsilon f(y,\dot{y},\ddot{y}) \\ f(y,-\dot{y},\ddot{y}) &= -f(y,\dot{y},\ddot{y}) \end{aligned}$$

are periodic.

Let us now describe another method for reducing autonomous differential equations to standard form (6-1). We consider only the special case

$$\begin{aligned} &\ddot{w}_j + \sigma_j^2 w_j = \epsilon f_j(w_1,\dot{w}_1, \ldots ,w_n,\dot{w}_n) \qquad j = 1, 2, \ldots , n \\ &\sigma_j \neq m\sigma_1 \qquad j = 2, 3, \ldots , n \qquad m = 0, \pm 1, \pm 2, \ldots \end{aligned} \tag{10-13}$$

The hypothesis on the σ_j in (10-13) ensures that the second-order equations $\ddot{w}_j + \sigma_j^2 w_j = 0$, $j = 2, 3, \ldots , n$, have no periodic solutions of period $2\pi/\sigma_1$.

The transformation $w_j = u_{2j-1}$, $\dot{w}_j = u_{2j}$,

$$\begin{aligned} u_1 &= x \sin \sigma_1\theta \\ u_2 &= \sigma_1 x \cos \sigma_1\theta \\ u_j &= y_{j-2} \qquad j = 3, 4, \ldots , 2n \end{aligned} \tag{10-14}$$

yields the equivalent system

$$\begin{aligned} \dot{\theta} &= 1 + \frac{\epsilon}{\sigma_1^2 x} f_1(x \sin \sigma_1\theta, \sigma_1 x \cos \sigma_1\theta, y_1, \ldots , y_{2n-2}) \sin \sigma_1\theta \\ \dot{x} &= \frac{\epsilon}{\sigma_1} f_1(x \sin \sigma_1\theta, \sigma_1 x \cos \sigma_1\theta, y_1, \ldots , y_{2n-2}) \cos \sigma_1\theta \\ \dot{y}_{2j-1} &= y_{2j} \\ \dot{y}_{2j} &= -\sigma_{j+1}^2 y_{2j-1} + \epsilon f_{j+1}(x \sin \sigma_1\theta, \sigma_1 x \cos \sigma_1\theta, y_1, \ldots , y_{2n-2}) \\ &\qquad\qquad j = 1, 2, \ldots , n - 1 \end{aligned} \tag{10-15}$$

Since the right-hand sides of (10-15) do not depend on t, one can eliminate t to obtain

$$\frac{dx}{d\theta} = \epsilon X(\theta,x,y)$$
$$\frac{dy}{d\theta} = B_2 y + \epsilon Y(\theta,x,y) \tag{10-16}$$

where $y = \text{col}\,(y_1, \ldots, y_{2n-2}) \qquad B = \text{diag}\,(C_2, \ldots, C_n)$

$$C_j = \begin{bmatrix} 0 & 1 \\ -\sigma_j^2 & 0 \end{bmatrix} \qquad j = 2, 3, \ldots, n \tag{10-17}$$

and X, Y are periodic in θ of period $2\pi/\sigma_1$. Since the σ_j satisfy (10-13), this is a special case of system (6-1) with $A = \text{diag}\,(0,B_2)$, and the results of Chap. 6 may be applied. In this manner, one obtains a periodic solution $x(\theta)$, $y(\theta)$, of (10-16) of period $2\pi/\sigma_1$ in θ. The period of the functions in (10-14) as a function of t are then determined by solving the equation

$$\dot{\theta} = 1 + \frac{\epsilon}{\sigma_1^2 x} f_1[x(\theta) \sin \sigma_1\theta, \sigma_1 x(\theta) \cos \sigma_1\theta, y(\theta)] \sin \sigma_1\theta$$

for $\theta = \theta(t,\epsilon)$, $\theta(0,\epsilon) = 0$ and choosing $\tau = \tau(\epsilon)$ so that $\theta(\tau,\epsilon) = 2\pi/\sigma_1$. It follows that τ is a period in t since $\theta(t + \tau, \epsilon) = \theta(t,\epsilon) + 2\pi/\sigma_1$ for all t.

The discussion above has shown how to obtain a periodic solution of a nonlinear autonomous differential equation. We now indicate how to decide whether or not such a periodic solution is stable. The analysis depends upon calculating the first approximation to the characteristic exponents and then using Theorem 4-2.

Consider the system of second-order equations

$$\ddot{w} + Aw = \epsilon f(w,\dot{w})$$
$$A = \text{diag}\,(\sigma_1^2, \ldots, \sigma_n^2) \tag{10-18}$$

where each $\sigma_j > 0$ and it is assumed that $f(w,\dot{w})$ is continuous together with its first and second partial derivatives with respect to w, $\dot{w}$ for all w, $\dot{w}$ in some region.

Furthermore, we suppose that

$$\sigma_j - \sigma_k \neq m\sigma_1 \qquad j \neq k \qquad j, k = 2, 3, \ldots, n \qquad m = 0, \pm 1, \pm 2, \ldots \tag{10-19}$$

and that there exists a periodic solution $w^*(t,\epsilon)$ of (10-18) of period $T = 2\pi/\omega(\epsilon)$, $\omega(0) = \sigma_1$, such that

$$w^*(t,0) = \text{col}\,(a \sin(\sigma_1 t + \varphi), 0, \ldots, 0) \qquad a, \varphi \text{ constant} \tag{10-20}$$

Theorem 10-1. If system (10-18) satisfies (10-19) and has a periodic solution $w^*(t,\epsilon)$ which satisfies (10-20), then, for ϵ sufficiently small, this periodic solution is asymptotically orbitally stable with asymptotic phase if

$$\epsilon \int_0^{2\pi/\sigma_1} f_{j\dot{w}_j}[w^*(t,0), \dot{w}^*(t,0)]\, dt < 0 \qquad j = 1, 2, \ldots, n$$

where $w^*(t,0)$ is defined in (10-20) and $f_{j\dot{w}_j}$ is the partial derivative of the jth coordinate of f with respect to the jth coordinate of $\dot{w}$.

Proof. The proof proceeds exactly as in the proof of Theorem 9-1. One first considers the linear variational equations (9-19) with respect to w^* for our system (10-18) and then applies Theorem 4-2. The characteristic exponents which reduce to $\pm i\sigma_j$ for $\epsilon = 0$, $j = 2, 3, \ldots, n$, are shown to have negative real parts exactly as in the proof of Theorem 9-1 and they are given explicitly by

$$\text{Re}\,(\tau_{2j-1}(\epsilon)) = \frac{\epsilon}{2T} \int_0^T f_{j\dot{w}_j}[w^*(t,0),\dot{w}^*(t,0)]\, dt \qquad j = 2, 3, \ldots, n \tag{10-21}$$

For $\epsilon = 0$, the other two characteristic multipliers are equal to 1, and, furthermore, since our system is autonomous, there is a periodic solution of the linear variational equations (as explained in Chap. 4) and one of these characteristic multipliers must be 1 for all ϵ. This implies $\tau_1(\epsilon)$, $\tau_2(0)$ may be taken equal to zero. Furthermore, from formula (3-2)

$$\tau_1 + \tau_2 + \tau_3 + \cdots + \tau_{2n} = \frac{\epsilon}{T} \int_0^T \sum_{j=1}^{n} f_{j\dot{w}_j}[w^*(t,\epsilon),\dot{w}^*(t,\epsilon)]\, dt$$

for all ϵ. Since $\tau_1 = 0$, it follows from (10-21) that, for ϵ sufficiently small, the first approximation to Re (τ_2) is given by

$$\text{Re}\,(\tau_2(\epsilon)) = \frac{\epsilon}{T} \int_0^T f_{1\dot{w}_1}[w^*(t,0),\dot{w}^*(t,0)]\, dt$$

and this is negative by the hypothesis of the theorem. Finally, $2n - 1$ of the characteristic exponents of the linear variational equations have negative real parts for ϵ sufficiently small and the theorem follows from Theorem 4-2.

As an exercise, show that the system

$$\begin{aligned} \ddot{x}_1 + \sigma^2 x_1 &= \epsilon(1 - x_1^2 - x_2^2)\dot{x}_1 \\ \ddot{x}_2 + 2\sigma^2 x_2 &= \epsilon(1 - x_1^2 - x_2^2)\dot{x}_2 \qquad \epsilon > 0 \end{aligned}$$

has two periodic solutions both of which are asymptotically orbitally stable with asymptotic phase.

11: Generalizations

To motivate what is to follow, let us briefly review the essential ideas of the method of Chap. 6 for obtaining periodic solutions of weakly nonlinear differential systems. More specifically, consider the system

$$\dot{y} = \epsilon q(t,y,\epsilon) \tag{11-1}$$

where ϵ is a real parameter and $q(t,y,\epsilon)$ is periodic in t of period $T = 2\pi/\omega$, $\omega > 0$, and sufficiently smooth in its arguments.

What we showed in Chap. 6 was that all the periodic solutions of (11-1) which for $\epsilon = 0$ reduce to a constant can be obtained in the following way: If a is any given constant vector, then there exists a periodic function $y^*(t,a,\epsilon)$ such that the mean value of $y^*(t,a,\epsilon)$ is a, and, furthermore, the derivative of y^* satisfies (11-1) except for a constant, and this constant is the negative of the mean value of $q(t,y^*(t,a,\epsilon),\epsilon)$. That is, all the Fourier coefficients of $y^*(t,a,\epsilon)$ have been determined in terms of its mean value a in such a way that all the Fourier coefficients of the function $\dot{y}^* - q(t,y^*,\epsilon)$ vanish except for the first. The first coefficient vanishes if the mean value of $q(t,y^*(t,a,\epsilon),\epsilon)$ vanishes; that is, if the determining equations (6-11) are satisfied. Finally, the function $y^*(t,a,\epsilon)$ was obtained by the method of successive approximations (6-9). This method of successive approximations indicates that $y^{(k+1)}(t)$ is obtained from the function $y^{(k)}(t)$ by first subtracting the mean value of $q(t,y^{(k)}(t),\epsilon)$ from $q(t,y^{(k)}(t),\epsilon)$ and then integrating the equation

$$y^{(k+1)}(t) = \epsilon(I - P_0)q(t,y^{(k)}(t),\epsilon) \tag{11-2}$$

for the unique function which has mean value a. Here, P_0q simply designates the projection of q onto its mean value and I is the identity.

Let us now enumerate the essential points which required investigation to make the above process rigorous. Recall that S is the space of continuous periodic functions of period T with norm ν, and define $S_1 \subset S$ to be the set of constant functions. For any $f \in S$, we defined the mean value of f to be $P_0 f$. Thus we see that P_0 is an operator which takes $S \to S$. With this notation, to validate the method of successive approximations (11-2) required the following:

1. If $f \in S$ is given, then there is a unique solution $y_0(t)$ of the equation

$$\dot{y} = f(t) \tag{11-3}$$

such that $y_0 \in S$, $P_0 y_0 = 0$, if and only if $P_0 f = 0$. Furthermore, if we define the operator H by $y_0 = Hf$, then there is a constant K independent of f such that

$$\nu(y_0) = \nu(Hf) \leqq K\nu(f) \tag{11-4}$$

2. If $f \in S$, then there exists an element $a_p \in S_1$ such that $P_0(f - a_p) = 0$.

3. For any given $a \in S_1$, $\|a\| \leqq b$, there exists a set $S_0 \subset S$, and $\epsilon_0 > 0$ such that, for any $f \in S_0$, $0 \leqq |\epsilon| \leqq \epsilon_0$,

$$y = \mathfrak{T}f = a + \epsilon Hqf \in S_0$$

$qf(t) = q(t,f(t),\epsilon)$ and this map is a contraction on S_0.

Properties 1 and 2 above showed that the method of successive approximations (11-2) was well defined, and property 3 was used in proving the convergence of the procedure. In Chap. 6, the set S_0 was defined to be the set of all $f \in S$ such that $P_0 f = a$, $\nu(f) \leqq d$.

We now turn to the determination of periodic solutions of systems which are different from (11-1) and orient our discussion in such a way as to make clear the relationship with the procedure applied before. It will be clear after one has read to the end of this chapter that many other problems can be solved by the same procedures.

11-1. ARBITRARY NONLINEAR SYSTEMS

Consider the system

$$\dot{y} = q(t,y) \tag{11-5}$$

where y, q are n vectors, q is a real function of t, y, is periodic in t of period $T = 2\pi/\omega$, is continuous in t, y, and has continuous derivatives with respect to y in $0 \leqq \|y\| \leqq R$, $-\infty < t < \infty$ (hypotheses this severe are not necessary; see Cesari [3]). Our problem is to obtain conditions on the function q which will ensure that there is a periodic solution of (11-5) of period T.

In the case where q contained a small parameter ϵ we determined a periodic function $y^*(t,a,\epsilon)$, with the mean value of y^* equal to a, in such a way that all the Fourier coefficients of the function $\dot{y}^* - q(t,y^*)$ vanished except for the first one. Now, if q does not contain a small parameter, it does not seem reasonable that, in general, the dominant term of the desired periodic function will be the mean value. In fact, there may be other harmonics in this function which are just as important as the mean value.

Cesari [3] has proposed the following procedure: Let

$$x_0(t) = \sum_{|R| \leqq m} a_k e^{ik\omega t} \tag{11-6}$$

be a given periodic function with all Fourier coefficients vanishing except the "first" $2m + 1$. Let us try to determine a periodic function $y^*(t,a) = x_0(t) + y_1^*(t,a)$, $a = (a_0, a_1, a_{-1}, \ldots, a_m, a_{-m})$, such that the first $2m + 1$ Fourier coefficients of $y_1^*(t,a)$ are equal to zero and all the Fourier coefficients of the function $\dot{y}^* - q(t,y^*)$ vanish *except the first* $2m + 1$. If this can be done, then necessary conditions for the existence of a periodic solution of (11-5) will be that the $(2m + 1)$ vector a be chosen in such a way as to make the *first* $(2m + 1)$ Fourier coefficients of $\dot{y}^* - q(t,y^*)$ vanish. These equations again will be referred to as the *determining equations*. If this procedure can be justified, then it is a direct generalization of the procedure above for the case where q contained a small parameter.

Let us now state the process more mathematically. Let S be the space of all real continuous periodic functions of period T with the L_2 norm ν; that is, if $f \in S$,

$$\nu(f) = \frac{1}{T} \int_0^T f'(t) f(t)\, dt \tag{11-7}$$

where f' is the transpose of f. If $f \in S$, then

$$f \sim \sum_{k=-\infty}^{+\infty} a_k e^{ik\omega t} \qquad a_k = \bar{a}_{-k} \qquad T = 2\pi/\omega \tag{11-8}$$

For a fixed integer $m \geqq 0$ and $f \in S$, define $P_m f$ to be

$$P_m f = \sum_{|k| \leqq m} a_k e^{ik\omega t} \tag{11-9}$$

and let

$$S_1 = \{f \in S \,|\, (I - P_m) f = 0\} \tag{11-10}$$

that is, S_1 is the set of periodic functions with the only possible nonzero Fourier coefficients being the first $2m + 1$, $a_0, a_1, a_{-1}, \ldots, a_m, a_{-m}$.

If $f \in S$ is any given function, then there is a unique solution $y_0(t)$ of the system $\dot{y} = f(t)$ with $y_0 \in S$, $P_m y_0 = 0$ if and only if $P_m f = 0$; more specifically, if $f \sim \sum_{|k|>m} a_k e^{ik\omega t}$, then

$$y_0 = Hf = \sum_{|k|>m} \frac{a_k}{ik\omega} e^{ik\omega t} \tag{11-11}$$

that is, H is the simple operation of taking the integral of f and neglecting all constant terms. If $f \in S$, $P_m f = 0$, then it is not difficult to see that

$$\begin{aligned} \nu(Hf) &\leqq \frac{1}{(m+1)\omega} \nu(f) \\ \|Hf\| &\leqq \frac{\sqrt{2}}{\omega} \sigma(m)\nu(f) \\ \sigma(m) &= \frac{1}{(m+1)^2} + \frac{1}{(m+2)^2} + \cdots \end{aligned} \tag{11-12}$$

Furthermore, for any $f \in S$, we have $P_m f \in S_1$ and $P_m[f - P_m f] = 0$. Consequently, for any $f \in S$, and any fixed $x_0 \in S_1$, the map

$$y = \mathfrak{F} f = x_0 + H(I - P_m)qf \qquad qf(t) = q(t,f(t)) \tag{11-13}$$

is well defined and $f \in S$, $P_m \mathfrak{F} f = x_0$. Now, for any given positive constants b, c, d, e, $b < d$, $c < e < R$, define the set $S_0 \subset S$ by the relation

$$S_0 = \{f \in S | P_m f = x_0, \nu(f) \leqq d, \|f\| \leqq e\} \tag{11-14}$$

where $x_0 \in S_1$ is given, $\nu(x_0) \leqq b$, $\|x_0\| \leqq c$. We define a method of successive approximations by the following relations:

$$\begin{aligned} y^{(0)} &= x_0 \\ y^{(k+1)}(t) &= \mathfrak{F} y^{(k)}(t) = x_0(t) + H(I - P_m)qy^{(k)}(t) \\ & \qquad k = 0, 1, 2, \ldots \end{aligned} \tag{11-15}$$

where H is the operator defined in (11-11).

From (11-12), it is now very easy to show that, if m is sufficiently large (a crude estimate of m is easy to obtain), then the sequence $\{y^{(k)}(t)\}$ defined by (11-15) converges to a function $y^*(t,a)$ with $y^* \in S_0$, and y^* is differentiable with respect to t, a. Consequently, from (11-15), the function $y^*(t,a)$ satisfies the equation

$$\dot{y}^*(t,a) = q(t,y^*(t,a)) + \dot{x}_0(t) - P_m q(t,y^*(t,a)) \qquad (11\text{-}16)$$

which is the same as (11-5) except for the term $\dot{x}_0(t) - P_m q(t,y^*)$. Finally, if the coefficients of the periodic function $x_0(t)$ (namely, the vector a) can be determined so that

$$\dot{x}_0(t) - P_m q(t,y^*(t,a)) = 0 \qquad \text{for all } t \qquad (11\text{-}17)$$

then the function $y^*(t,a)$ will be a periodic solution of (11-5). Equations (11-17) are referred to as the *determining equations.*

Of course, a discussion of equations of the form (11-17) is extremely difficult. One manner in which to proceed toward the solution of (11-17) is the following: Choose for the first approximation for the vector a that vector a_0 (if it exists) which is defined by the system of equations

$$\dot{x}_0(t) - P_m q(t,x_0(t)) = 0 \qquad \text{for all } t \qquad (11\text{-}18)$$

A little thought shows one that equations (11-18) represent those equations which would be obtained if one substituted the function $x_0(t)$ in (11-5) and equated to zero the Fourier coefficients corresponding to the first m harmonics; i.e., these are the equations that would be obtained by applying the Ritz-Galerkin procedure. If one can show that equations (11-18) having a solution implies that equations (11-17) have a solution then a periodic solution of (11-5) exists. The verification of this fact is extremely difficult but has been carried out in some specific cases by Cesari [3]. Knobloch [1] recently has shown that the uniform topology may replace the use of the L^2 norm in this process and has given further applications of this method.

11-2. EQUIVALENT LINEARIZATION AND DESCRIBING FUNCTIONS

In this section, we discuss how the procedure discussed in Sec. 11-1 may be useful for the determination of the range of validity of the methods of equivalent linearization and describing functions. Suppose we have a system

$$\dot{x} = f(x) \qquad f(0) = 0 \qquad (11\text{-}19)$$

where x, f are n vectors, and we wish to determine whether or not this equation has a periodic solution of some period ω. Let A be a given constant matrix such that all the solutions of the equation

$$\dot{x} = Ax \qquad (11\text{-}20)$$

are periodic of period 2π. We pose the following problem: What conditions on ω and the function $f(x)$ will ensure that system (11-19) has a periodic solution $x(t,\omega)$ of period $2\pi/\omega$ with

$$x(t,\omega) = e^{A\omega t}[x_0 + x_1(t)] \qquad \int_0^{2\pi/\omega} x_1(t)\,dt = 0 \tag{11-21}$$

where x_0 is a constant vector? In applications, it is desirable to be able to answer this question by determining a set of equations which involve only x_0 and ω such that, if x_0 and ω are solutions of these equations, then system (11-19) has a periodic solution of the form (11-21).

In such a situation, it is natural to introduce the transformation of variables

$$x = e^{A\omega t}y \tag{11-22}$$

and try to determine a periodic function y of period $2\pi/\omega$ with mean value equal to x_0 so that $x = e^{A\omega t}y$ is a periodic solution of (11-19). The differential equation for y is

$$\dot{y} = e^{-A\omega t}f(e^{A\omega t}y) - A\omega y \tag{11-23}$$

To analyze this equation, it is convenient to introduce a new time variable $\omega t = \tau$ to obtain

$$\frac{dy}{d\tau} = \frac{1}{\omega}\, e^{-A\tau}f(e^{A\tau}y) - Ay \overset{\text{def}}{\equiv} \frac{1}{\omega}\, B(\tau,y) - Ay \tag{11-24}$$

Now, the problem is to determine a periodic solution $y(\tau,\omega,x_0)$ of (11-24) of period 2π such that

$$\frac{1}{2\pi}\int_0^{2\pi} y(\tau,\omega,x_0)\,d\tau = x_0$$

and to obtain the existence of this solution by specifying auxiliary conditions only on x_0 and ω. Since (11-24) is a special case of (11-5), our problem will be solved if we can justify the procedure outlined for system (11-5) with $m = 0$. More precisely, we must show the following:

1. For $m = 0$, the method of successive approximations (11-15) converges to a function $y^*(\tau,\omega,x_0)$ for given $x_0 \in U$, $\omega \in E$, where U, E are given sets.

2. For $m = 0$, the determining equations (11-17) are satisfied. In this particular case, equations (11-17) are given by

$$\frac{1}{2\pi\omega}\int_0^{2\pi}[e^{-A\tau}f(e^{A\tau}y^*(\tau,x_0,\omega))]\,d\tau - Ax_0 = 0 \tag{11-25}$$

The verification of steps 1 and 2 is in general very difficult. If one supposes step 1 has been verified, then one must show that equations (11-25) have a solution where the function $y^*(\tau,x_0,\omega)$ can be determined only approximately. As mentioned in Sec. 11-1, one attempts to solve (11-25) by replacing $y^*(\tau,x_0,\omega)$ by the "first approximation" x_0 to obtain

$$\frac{1}{2\pi\omega}\int_0^{2\pi}e^{-At}f(e^{At}x_0)\,dt - Ax_0 = 0 \tag{11-26}$$

If equations (11-26) have a solution x_0, ω, then it is clear what has to be proved in order to show that system (11-19) has a periodic solution.

Let us give another interpretation of equations (11-25). If one defines

$$\begin{aligned} B_0(y) &\overset{\text{def}}{=} \frac{1}{2\pi}\int_0^{2\pi}B(\tau,y)\,d\tau \\ &= \frac{1}{2\pi}\int_0^{2\pi}e^{-A\tau}f(e^{A\tau}y)\,d\tau \end{aligned} \tag{11-27}$$

and averages the right-hand side of (11-24) over a period of τ, one obtains the "averaged" equations

$$\frac{dy}{d\tau} = \frac{1}{\omega}B_0(y) - Ay \tag{11-28}$$

Notice that the right-hand side of (11-28) is the same as the left-hand side of (11-26). Consequently, if there are values ω_0 and x_0 such that, for $\omega = \omega_0$, the constant function $y = x_0$ is a solution of (11-28), then ω_0, x_0 satisfy (11-26) and thus there is an "approximate" periodic solution of (11-19), $x = [\exp(A\omega_0 t)]x_0$, with the "approximate" period $2\pi/\omega_0$. An analysis of the other solutions of (11-28) will yield the "approximate" transient behavior of the solutions of (11-19). The word "approximate" is always written with quotation marks since many things must be proved to justify the process.

The method of equivalent linearization involves a study of the linear variational equations associated with the equilibrium point ω_0, x_0 of (11-28), and the method of describing functions is a graphical procedure for analyzing the behavior of the solutions of (11-28). Bass [1,2] and Glatenok [1] have discussed the validity of these procedures.

11-3. ALMOST LINEAR SYSTEMS

Consider the system

$$\dot{y} = A(t)y + \epsilon q(t,y,\epsilon) \tag{11-29}$$

where y, q are n vectors, ϵ is a small real parameter, A is an $n \times n$ matrix, A, q are periodic in t of period T, are continuous in t, y, ϵ, and have continuous first derivatives with respect to y, ϵ for all $t \geqq 0$, $0 \leqq \|y\| \leqq R$, $R > 0$, $0 \leqq |\epsilon| \leqq \epsilon$, $\epsilon_0 > 0$. Since $A(t)$ is periodic in t, there exists a periodic change of variables which would reduce (11-29) to a system of the form (9-1) and the theory of Chap. 9 could be applied. However, we prefer to treat (11-29) directly to show that the methods other authors have proposed for obtaining periodic solutions for systems of the form (11-29) (see, for example, Malkin [1], Friedrichs [1], Lewis [1,2], and Sibuya [1]) use essentially the same ideas as above.

To obtain a periodic solution of system (11-29) by successive approximations, it is first necessary to know when the linear nonhomogeneous equation

$$\dot{y} = A(t)y + f(t) \tag{11-30}$$

has a periodic solution of period T when f is periodic of period T.

Again let S be the space of all periodic n-vector functions of period T with norm ν, $\nu(f) = \sup_{0 \leqq t \leqq T} \|f(t)\|$. Let $\varphi_j \in S$, $j = 1, 2, \ldots, k$, be a basis for the periodic solutions of the homogeneous equation (11-30); let $\Phi = (\varphi_1, \ldots, \varphi_k)$; let $\psi_j \in S$, $j = 1, 2, \ldots, k$, be a basis for the periodic solutions of the adjoint equation

$$\dot{z} = -A'(t)z \qquad (A' \text{ is the transpose of } A) \tag{11-31}$$

and let $\Psi = (\psi_1, \ldots, \psi_k)$. Define

$$\begin{aligned} S_1 &= \{f \in S | f = \Phi a, a \in E^k\} \\ S_2 &= \{f \in S | f = \Psi a, a \in E^k\} \end{aligned} \tag{11-32}$$

and define operators P, Q such that $P^2 = P$, $Q^2 = Q$, the identity, such that P, Q are continuous on S and $PS \subset S_1$, $QS \subset S_2$, and if $f \in S_1$, $Pf = f$, $f \in S_2$, $Qf = f$; that is

$$\begin{aligned} &P \text{ is a projection operator projecting } S \text{ onto } S_1 \\ &Q \text{ is a projection operator projecting } S \text{ onto } S_2 \end{aligned} \tag{11-33}$$

The sets S_1, S_2 consist of all periodic solutions of the homogeneous part of (11-30) and the adjoint equation (11-31), respectively.

We now digress for a moment and prove a basic lemma for the existence of periodic solutions of (11-30).

Lemma 11-1. There is a unique solution $y_0(t)$ of (11-30) with $y_0 \in S$, $Py_0 = 0$ if and only if $Qf = 0$. Furthermore, if we define the operator H by $y_0 = Hf$, then there is a constant K independent of f such that

$$\nu(y_0) = \nu(Hf) \leqq K\nu(f)$$

Proof. If $Z(t)$, $Z(0) = I$, is a fundamental system of solutions of $\dot{z} = A(t)z$, then the solution of (11-30) with $y(0) = y_0$ is

$$y(t) = Z(t)y_0 + \int_0^t Z(t)Z^{-1}(\tau)f(\tau)\,d\tau$$

This function $y(t)$ will be periodic of period T if and only if y_0 satisfies the system of linear equations

$$Dy_0 = b \tag{11-33a}$$

where $\quad D = [Z(T) - I] \qquad b = -\int_0^T Z^{-1}(\tau)f(\tau)\,d\tau$

If we let $\mathcal{R}(D)$ and $\mathfrak{N}(D)$ designate the range of D and the null space of D, respectively, then a fundamental relation in the theory of matrices is that $\mathcal{R}(D) \oplus \mathfrak{N}(D') = E^n$, where D' is the transpose of D. A necessary and sufficient condition that (11-33a) has a solution is that $b \in \mathcal{R}(D)$ or, equivalently, that the projection of b onto $\mathfrak{N}(D')$ is zero. If C is a continuous operation such that $Cx = x$, $x \in \mathfrak{N}(D')$ and $C(E^n) \subset \mathfrak{N}(D')$ then (11-33a) has a solution y_0 if and only if $Cb = 0$. Furthermore, if $Cb = 0$, and $z_0 \in \mathfrak{N}(D)$ is a given vector, then $\mathcal{R}(D') \oplus \mathfrak{N}(D)$ implies (11-33a) has a unique solution $y_0 = z_0 + y_0^*$, $y_0^* \in \mathcal{R}(D')$, y_0^* independent of z_0, and there exists a constant K independent of b such that $\|y_0^*\| \leqq K\|b\|$. It remains to interpret these results in terms of periodic solutions of $\dot{z} = A(t)z$ and the adjoint equation (11-31). It is clear that the set of vectors $z_0 \in \mathfrak{N}(D)$ coincides with the set of initial values which lead to periodic solutions of $\dot{z} = A(t)z$. Also the set of $z_0 \in \mathfrak{N}(D')$ coincides with the set of initial values which lead to periodic solutions of the adjoint equation (11-31). The remainder of the proof of the lemma follows immediately.

To proceed further, we need some more notation. If f, $g \in S$, let

$$(f,g) = \frac{1}{T}\int_0^T f'(t)g(t)\,dt$$

If $f \in S$, $R_{1f} = \text{col}\,[(f,\varphi_1), \ldots, (f,\varphi_k)]$, $R_{2f} = \text{col}\,[(f,\psi_1), \ldots, (f,\psi_k)]$, then $Pf = \Phi R_{1f}$, $Qf = \Psi R_{2f}$. It is not too difficult to show that the matrix $R = (R_{1,\psi_j}, j = 1, 2, \ldots, k) = (R_{2,\varphi_j}, j = 1, 2, \ldots, k)'$ is nonsingular. From this fact, it follows that, for any $f \in S$, there exists a unique vector a_f such that $Q(f - \Psi a_f) = 0$. In fact, if $b \in E^k$, $z = \Psi b$ then $Q(f - \Psi b) = Qf - \Psi R'b = \Psi[R_{2f} - R'b]$. Consequently, this is zero if and only if $R'b = R_{2f}$, which implies b is unique.

With these preliminaries we can now define our method of successive approximations as follows:

$$\begin{aligned} y^{(0)} &= \Phi a \qquad a \in E^k, \text{ and fixed} \\ y^{(k+1)} &= \Phi a + \epsilon H[q(t,y^{(k)}(t),\epsilon) - \Psi a_{q^{(k)}}] \qquad k = 0, 1, \ldots \end{aligned} \tag{11-34}$$

where $q^{(k)} = q(t,y^{(k)}(t),\epsilon) \qquad a_{q^{(k)}} = (R')^{-1}R_{2q^{(k)}} \qquad k = 0, 1, 2, \ldots$

It is easy to show that for any given vector a such that $\nu(\Phi a) \leqq b < R$, there exists an $\epsilon_1 > 0$ such that the method of successive approximations (11-34) converges for $0 \leqq |\epsilon| \leqq \epsilon_1$ to a function $y^*(t,a)$ which belongs to S and

$$\dot{y}^* = A(t)y^* + \epsilon q(t,y^*,\epsilon) - \epsilon\Psi(t)(R')^{-1}R_{2,q(t,y^*,\epsilon)} \tag{11-35}$$

where
$$R_{2,q(t,y^*,\epsilon)} = \frac{1}{T}\int_0^T \Psi'(t)q(t,y^*,\epsilon)\,dt$$

and R is a constant nonsingular matrix. Consequently, the function $y^*(t,a,\epsilon)$ will be a periodic solution of (11-29) if the *determining equations*

$$R_{2,q(t,y^*,\epsilon)} = 0$$

for some $a = a(\epsilon)$, $0 \leqq |\epsilon| \leqq \epsilon_2$, with $\nu(\Psi a(\epsilon)) \leqq b$.

It is sometimes necessary to investigate systems more general than (11-29). In fact, one must consider equations of the form

$$\dot{y} = A(t)y + q(t,y,\epsilon) \tag{11-36}$$

where $A(t)$, $q(t,y,\epsilon)$ are periodic in t of period T and

$$q(t,y,\epsilon) = O(\|y\|^2 + |\epsilon|)$$

as $\|y\|$, $|\epsilon| \to 0$. In this case, one wishes to determine conditions for the existence of a periodic solution which approaches zero as $\epsilon \to 0$. For problems of this type, the matrix A and the vector q generally contain some parameters which are to be chosen in such a way as to obtain a periodic solution of (11-36). If this is the case, then it is clear that the above method of successive approximations (11-34) can be applied by simply taking the initial approximation $y^{(0)}$ as $y^{(0)} = 0$.

The parameters in A, q are then chosen so that the determining equations are satisfied.

An example where one needs this more general procedure is the following: Suppose one knows a family of periodic solutions

$$x = x^0[\omega(b)t + \varphi, b] \qquad b, \varphi \text{ constants}$$
$$x^0(s + 2\pi, b) = x^0(s,b)$$

of the autonomous system

$$\dot{x} = f(x)$$

and one wishes to find conditions on b, φ which will ensure that, for ϵ sufficiently small, the system

$$\dot{x} = f(x) + \epsilon F(t,x) \qquad F(t + T_1, x) = F(t,x)$$

has a periodic solution of period T where T is some integral multiple of T_1. Furthermore, one wants this periodic solution for $\epsilon = 0$ to coincide with $x^0[\omega(b)t + \varphi, b]$ for some b, φ. Suppose that b is such that $[2\pi/\omega(b)]/T_1$ is rational and let T be the least common multiple of $2\pi/\omega(b)$ and T_1. If for this b, $x = x^0[\omega(b)t + \varphi, b] + y$, then

$$\dot{y} = A(t)y + q(t,y,\epsilon)$$

where $A(t)$, $q(t,y,\epsilon)$ are periodic in t of period T, $q(t,y,\epsilon) = O(\|y\|^2 + |\epsilon|)$, as $\|y\|$, $|\epsilon| \to 0$ and

$$A(t) = \frac{\partial f\,\{x^0[\omega(b)t + \varphi, b]\}}{\partial x}$$

This is now a special case of (11-36) and the above theory may be applied. Notice that the number of characteristic multipliers of our linear system which are equal to 1 is given by 1 + dimension of b, since all the functions $\partial x^0/\partial b$, $\partial x^0/\partial \varphi$ are solutions of this linear system. To obtain sufficient conditions on b, φ which will ensure the existence of periodic solutions is extremely complicated and will not be given. The interested reader may consult the papers of Loud [1,2] where these conditions are given for the case where b is a scalar. The computations of Loud were carried out by using essentially the perturbation procedure of Poincaré.

11-4. HIGHER-ORDER EQUATIONS

Consider the scalar equation

$$\frac{d^n y}{dt^n} = \epsilon Q(t,y,\epsilon) \tag{11-37}$$

where ϵ is a small real parameter, $q(t,y,\epsilon)$ is a sufficiently smooth function of t, y, ϵ, is periodic in t of period T. We have already indicated in Chap. 9 and Sec. 7-8 how to obtain periodic solutions of (11-37) of period T. However, in so doing, the solution may depend upon fractional powers of the parameter ϵ. By a detailed discussion of the determining equations, one could determine the dependence of the solution upon ϵ by use of the Newton polygon (see, for example, Imaz [1], Kushul [1]).

However, it may be desirable to obtain directly sufficient conditions on the function q which ensure that (11-37) has a periodic solution which is analytic in ϵ if q is analytic in y, ϵ in the neighborhood of $y = 0$, $\epsilon = 0$. A method to accomplish this has been given by Bogoliubov and Sadovnikov [1]. As we shall see, it is precisely the same type of procedure that we have used before.

Let S be the space of periodic functions of period T with norm ν, $\nu(f) = \sup_{0 \leq t \leq T} |f(t)|$. Let S_1 be the subset of S which consists of all constant functions, and if $f \in S$, let $P_0 f$ be the mean value of f.

If $f \in S$, then there is a unique periodic solution $y_0(t)$ of the equation

$$\frac{d^n y}{dt^n} = f(t)$$

with $y_0 \in S$, $P_0 y_0 = 0$, if and only if $P_0 f = 0$. Furthermore, if we let $y_0 = Hf$, then

$$\nu(y_0) = \nu(Hf) \leq K\nu(f)$$

for some constant K independent of f.

The remainder of the discussion is exactly the same as in Chap. 6. The method of successive approximations is

$$\begin{aligned} y^{(0)} &= a \qquad a \text{ fixed} \\ y^{(k+1)} &= a + \epsilon H(I - P_0)q(t,y^{(k)}(t),\epsilon) \qquad k = 0, 1, 2, \ldots \end{aligned} \tag{11-38}$$

For ϵ small, the process converges to a function $y^*(t,a,\epsilon)$, $y^* \in S$ and $y^*(t,a,\epsilon)$ will be a periodic solution of (11-37) if a can be chosen so that

$$\int_0^T q(t,y^*(t,a,\epsilon), \epsilon)\, dt = 0 \tag{11-39}$$

The method of Cesari given in Sec. 11-1 can easily be generalized to equation (11-27) without the small parameter ϵ.

11-5. GENERALIZED FLOQUET THEORY

Let Ω be a set of complex numbers, $\mathcal{B}$ be a subset of the complex plane which contains at least two points. Let Ω_0 be the set of complex numbers obtained by taking all finite linear combinations with non-negative integer coefficients of elements of Ω. Let $\mathcal{P}(\Omega,\mathcal{B})$ [or $\mathcal{P}(\Omega_0,\mathcal{B})$] be the set of functions $f(z)$ of the complex variable z of the form

$$f(z) = \Sigma a_k \exp \omega_k z \qquad [\text{or } f(z) = \Sigma a_k \exp \omega_k^0 z] \tag{11-40}$$

where $\omega_k \in \Omega$ [or $\omega_k^0 \in \Omega_0$] and the series are absolutely convergent uniformly in $\mathcal{B}$.

If $A(z)$ is an $n \times n$ matrix whose elements all belong to $\mathcal{P}(\Omega,\mathcal{B})$, A_0 is an $n \times n$ constant matrix, and λ is a complex parameter, we wish to consider the equation

$$\frac{dw}{dz} = A_0 w + \lambda A(z) w \tag{11-41}$$

Golomb [3] has discussed the following question for systems (11-41): Under what conditions is there a fundamental matrix W of solutions of (11-41) of the form

$$W = U(z)e^{Tz} \tag{11-42}$$

where each element of $U(z)$ is in $\mathcal{P}(\Omega_0,\mathcal{B})$ and T is a constant matrix?

For $\lambda = 0$, a fundamental system of (11-41) is $W = \exp(A_0 z)$. Consequently, one would want to choose $U(z)$, T in such a way that $\|U(z) - I\| \to 0$ as $\lambda \to 0$, $\|T - A_0\| \to 0$ as $\lambda \to 0$.

If such a fundamental system (11-42) exists, we call the eigenvalues of the matrix T *generalized characteristic exponents* of (11-41). This generalizes the usual concept of characteristic exponents since, if Ω is the set of complex numbers of the form $\pm in\omega$, $\omega > 0$, $n = 0, 1, 2, \ldots$, then $\Omega_0 = \Omega$, the functions $f \in \mathcal{P}(\Omega,\mathcal{B})$ are periodic functions, and the representation (11-42) is the usual Floquet representation of Theorem 3-1.

We wish to show that the results obtained by Golomb [3] for system (11-41) can be derived from the ideas we have been using above.

Let W be an $n \times n$ matrix and consider the matrix equation

$$\frac{dW}{dz} = A_0 W + \lambda A(z) W \tag{11-43}$$

where λ, A_0, $A(z)$ are the same as above. Let T_1 be any given $n \times n$ constant matrix and consider the transformation

$$W = U \exp (A_0 - T_1)z \tag{11-44}$$

from W to U. The new equation for U is

$$\frac{dU}{dz} = A_0 U - U A_0 + U T_1 + \lambda A(z) U \tag{11-45}$$

If we can determine the matrix T_1 in such a way that (11-45) has a solution $U \in \mathcal{P}(\Omega_0,\mathcal{B})$ then our original problem will be solved for $T = A_0 - T_1$. For any $f \in \mathcal{P}(\Omega_0,\mathcal{B})$ define P_0f to be the constant term in the exponential expansion of f. If F is an $n \times n$ matrix with each of its elements in $\mathcal{P}(\Omega_0,\mathcal{B})$, then P_0F has the obvious interpretation. Since we wish to use a procedure analogous to that used in the previous sections, we need the following lemma.

Lemma 11-2. If $\sigma_1, \ldots, \sigma_n$ are the eigenvalues of A_0; if there exists a $\delta > 0$ such that

$$|\sigma_j - \sigma_k + \omega| \geqq \delta > 0 \qquad \text{for all } \omega \in \Omega_0 \qquad \omega \neq 0 \qquad j, k = 1, 2, \ldots, n \tag{11-46}$$

and if $F(z)$ is any given $n \times n$ matrix function $F \in \mathcal{P}(\Omega_0,\mathcal{B})$, $P_0F = 0$, then there is a solution $U(z)$ of the equation

$$\frac{dU}{dz} = A_0U - UA_0 + F(z) \tag{11-47}$$

with $U(z) = I + U_1(z)$, $U(z) \in \mathcal{P}(\Omega_0,\mathcal{B})$, where I is the identity and $\|U_1(z)\| \leqq K\|F(z)\|$ where K is a constant which does not depend on F. If the expansion of functions in $\mathcal{P}(\Omega_0,\mathcal{B})$ in a series of the form (11-40) is unique then the function $U(z)$ is the only solution of (11-47) with $P_0U = I$.

Proof. We prove this lemma by considering an exponential expansion of U. If

$$F(z) = \Sigma B_k \exp \omega_k z$$
$$U(z) = \Sigma C_k \exp \omega_k z \qquad \omega_k \in \Omega_0$$

and if the matrices C_k are solutions of the equation

$$\omega_k C_k - A_0C_k + C_kA_0 = B_k \tag{11-48}$$

for all k, then $U(z)$ will be a solution of (11-47). System (11-48) represents a system of linear equations for the elements of the matrices C_k, and the eigenvalues of the coefficient matrix of the linear system are $\omega_k - \sigma_j + \sigma_l$, $j, l = 1, 2, \ldots, n$. From the hypotheses, if $\omega_k \neq 0$ then the eigenvalues of the linear system (11-48) are bounded away from zero for all k and thus there is always a unique solution of (11-48). If $\omega_k = 0$, then $B_k = 0$ and any matrix C_k which commutes with A_0 satisfies (11-48); in particular, we can take the identity I. From the absolute convergence of the expansions of F uniformly in $\mathcal{B}$, the existence of $U(z)$ of the above type is proved. The uniqueness of the exponential expansions of functions in $\mathcal{P}(\Omega_0,\mathcal{B})$ implies there is only one way to obtain the equations (11-48) and the remainder of the lemma follows.

This lemma allows us to prove very easily the following theorem.

Theorem 11-1. If (11-46) is satisfied then there exist a number $\lambda_0 > 0$ and matrices $T(\lambda)$, $U(z,\lambda)$, $0 \leqq |\lambda| \leqq \lambda_0$, continuous in λ at $\lambda = 0$, $T(0) = A_0$, $U(z,0) = I$, $T(\lambda)$ constant, $U(z,\lambda) \in \mathcal{P}(\Omega_0,\mathcal{B})$ $dU/dz \in \mathcal{P}(\Omega_0,\mathcal{B})$ such that

$$W(z) = U(z,\lambda)e^{T(\lambda)z}$$

is a fundamental system of solutions of (11-41). If the exponential expansions of functions in $\mathcal{P}(\Omega_0,\mathcal{B})$ are unique then the generalized characteristic exponents are unique mod Ω_0.

Proof. Define the method of successive approximations $U^{(0)} = I$,

$$\frac{dU^{(k+1)}}{dz} = A_0U^{(k+1)} - U^{(k+1)}A_0 + (I - P_0)[U^{(k)}T_1 + \lambda A(z)U^{(k)}]$$
$$k = 0, 1, 2, \ldots$$

where $U^{(k)}$ is determined successively so that $U^{(k)} = I + U_1^{(k)}(z)$ and $U_1^{(k)}(z)$ is given in the above lemma. Then it is not difficult to show that, if $\|T_1\|$, $|\lambda|$ are sufficiently small, the sequence $\{U^{(k)}\}$ converges to a function $U^*(z,T_1,\lambda)$ which satisfies (11-45) except for the term $P_0[U^*T_1 + \lambda A(z)U^*]$ and U^* is continuous in λ at $\lambda = 0$,

$$U^*(z,T_1,0) = I$$

Consequently, from the implicit-function theorem, it follows that for λ sufficiently small one can always choose T_1 so that

$$P_0[U^*T_1 + \lambda A(z)U^*] = 0$$

and the first part of the theorem is proved. The last part of the theorem is an immediate consequence of the lemma.

For application of these results, especially to expansions near irregular singular points, see the paper of Golomb [3].

Part III

Almost Periodic Solutions and Integral Manifolds

12: Almost Periodic Functions and Multiply Periodic Functions

The purpose of the present chapter is to define and state some basic properties of almost periodic functions and multiply periodic functions. Approximation theorems are also given for the integrals of almost periodic functions of average zero. Some general references for almost periodic functions are Bohr [1], Besicovitch [1], and Favard [1].

Definition 12-1. A function $f(t,x)$, where f is an m vector, t is a real scalar, x is an n vector is said to be *almost periodic in t uniformly in x, $x \in \Lambda$*, if $f(t,x)$ is continuous in t, x for $t \in E$, $x \in \Lambda$ and if, for any $\eta > 0$, it is possible to find an $l(\eta) > 0$ such that, in any interval of length $l(\eta)$, there is a τ such that the inequality

$$\|f(t + \tau, x) - f(t,x)\| \leqq \eta$$

is satisfied for all $t \in E$, $x \in \Lambda$.

The following criterion of Bochner will be very useful: If Λ is compact, then $f(t,x)$ is almost periodic in t uniformly in x, $x \in \Lambda$, if and only if, for every sequence of real numbers $\{\tau_n\}$, the sequence $\{f(t + \tau_n, x)\}$ has a subsequence $\{f(t + \tau_{n_k}, x)\}$ which converges uniformly with respect to $t \in E$, $x \in \Lambda$.

The class of almost periodic functions of the above type contains all functions of the form

$$\Sigma a_j(x) \cos \omega_j t + b_j(x) \sin \omega_j t$$

where the sum is finite, the ω_j are given constants, and $a_j(x)$, $b_j(x)$ are given continuous functions of x for $x \in \Lambda$.

If $f(t,x)$ is almost periodic in t uniformly with respect to $x \in \Lambda, \Lambda$ is compact, $\varphi(t)$ is almost periodic in t and $\varphi(t) \in \Lambda$ for all t, then $f(t,\varphi(t))$ is almost periodic in t.

If $f(t,x)$ is almost periodic in t uniformly with respect to $x \in \Lambda$, then let $\{\lambda_j\}$ be the set of all real numbers such that

$$\lim_{T\to\infty} \frac{1}{T} \int_0^T f(t+\tau, x) \exp(-i\lambda_j\tau)\, d\tau$$

is not identically zero for $x \in \Lambda$. It is known that $\{\lambda_j\}$ is countable. The set $\{\omega_\alpha\}$ is called a set of *basic frequencies* for $f(t,x)$ if $\{\omega_\alpha\} \subset \{\lambda_j\}$ and

1. The ω_α are linearly independent over the rationals; that is, there exists no relation of the form $\Sigma_\alpha n_\alpha \omega_\alpha = 0$, where the sum is finite and n_α are rationals, unless all $n_\alpha = 0$.
2. Any λ_j in the set $\{\lambda_j\}$ defined above can be obtained as a finite linear combination of some of the ω_α with integer coefficients.

Notice that two different sets of basic frequencies have the same cardinal number.

A set of basic frequencies $\{\omega_\alpha\}$ satisfies the following properties: If $\{\tau_m\}$ is a sequence of real numbers such that, for each ω_α,

$$\exp i\omega_\alpha\tau_m \to 1 \text{ as } m \to \infty \tag{12-1}$$

then

$$\|f(t+\tau_m, x) - f(t,x)\| \to 0 \text{ as } m \to \infty \tag{12-2}$$

uniformly for $t \in E$, $x \in \Lambda$. On the other hand, if $\{\omega_\alpha\}$ is any countable sequence of real numbers, linearly independent over the rationals and if for each sequence of real numbers τ_m satisfying (12-1) one can prove that (12-2) is satisfied uniformly for $t \in E$, $x \in \Lambda$, then $f(t,x)$ will be almost periodic in t uniformly in x, $x \in \Lambda$ and $\{\omega_\alpha\}$ is a set of basic frequencies of $f(t,x)$.

Finally, for functions $f(t,x)$ which are almost periodic in t uniformly in x, $x \in \Lambda$, the following limit exists uniformly in t, x and is independent of t:

$$\lim_{T\to\infty} T^{-1} \int_0^T f(t+u, x)\, du \tag{12-3}$$

Here, and in the following, we shall call this limit the mean value of f with respect to t and designate it by

$$f_0(x) = \lim_{T\to\infty} T^{-1} \int_0^T f(t,x)\, dt \tag{12-4}$$

For periodic functions $f(t,x)$, this is the usual mean value.

The following approximation lemmas are due to Bogoliubov and Mitropolski [1].

Lemma 12-1. Suppose that $f(t,x)$, f an m vector, t a scalar, x an n vector is almost periodic in t uniformly in x, $x \in \Lambda$, a compact subset of E^n. If the mean value of $f(t,x)$ with respect to t is zero, that is, $f_0(x) = 0$, then, for every $\eta > 0$, there exists a continuous function $\zeta(\eta)$, $\zeta(\eta) \to 0$ as $\eta \to 0$ such that if

$$f_\eta(t,x) = \int_{-\infty}^{t} e^{-\eta(t-\tau)} f(\tau,x)\, d\tau \tag{12-5}$$

then

$$\|f_\eta(t,x)\| \leqq \eta^{-1}\zeta(\eta) \tag{12-6}$$

for all $t \in E$, $x \in \Lambda$. Finally,

$$\left\| \frac{\partial f_\eta(t,x)}{\partial t} - f(t,x) \right\| \leqq \zeta(\eta) \tag{12-7}$$

for all $t \in E$, $x \in \Lambda$.

Proof. There exists a continuous decreasing function $\epsilon(T)$, $\epsilon(T) \to 0$ as $T \to \infty$ such that $\left\| T^{-1} \int_t^{t+T} f(\tau,x)\, d\tau \right\| \leqq \epsilon(T)$ for all $t \in E$, $x \in \Lambda$. Furthermore,

$$f_\eta(t,x) = \sum_{n=0}^{\infty} e^{-\eta nT} \int_{nT}^{(n+1)T} f(t-\tau, x) e^{-\eta(\tau-nT)}\, d\tau$$

From the assumptions of the lemma on f, there exists a constant B such that $\|f(t,x)\| \leqq B$, $t \in E$, $x \in \Lambda$, and

$$\begin{aligned}
\|f_\eta(t,x)\| \leqq \sum_{n=0}^{\infty} e^{-\eta nT} \left\| \int_{nT}^{(n+1)T} f(t-\tau, x)\, d\tau \right\| \\
+ B \sum_{n=0}^{\infty} e^{-\eta nT} \int_{nT}^{(n+1)T} (1 - e^{-\eta(\tau-nT)})\, d\tau \\
\leqq \sum_{n=0}^{\infty} e^{-\eta nT}\epsilon(T)T + BT = T\epsilon(T)[1 - e^{-\eta T}]^{-1} + BT
\end{aligned}$$

Now choose T as a function of η so that $1 - e^{-\eta T} = \epsilon(T)$. Since $\epsilon(T) \to 0$ as $T \to \infty$, it is clear that the solution T_η of this equation satisfies $\eta T_\eta \to 0$ as $\eta \to 0$ and is continuous in η. Notice that $T_\eta \to \infty$ as $\eta \to 0$. If we let $(B+1)\eta T_\eta = \zeta(\eta)$, then the first part of the lemma is proved.

The relation (12-7) follows from the identity

$$\frac{\partial f_\eta(t,x)}{\partial t} - f(t,x) = -\eta f_\eta(t,x)$$

and (12-6).

Lemma 12-2. Suppose that $f(t,\theta,x,\epsilon)$, f a k vector, t a scalar, θ an m vector, x an n vector, ϵ a scalar, is a continuous function of t, θ, x, ϵ for $t \in E$, $\theta \in E^m$, $0 \leqq \|x\| \leqq \sigma$, $\sigma > 0$, $0 \leqq \epsilon \leqq \epsilon_0$, $\epsilon_0 > 0$, has continuous second derivatives with respect to θ, x, is multiply periodic* in θ with vector period ω, and for each fixed ϵ, is almost periodic in t uniformly with respect to θ, x, $\theta \in E^m$, $0 \leqq \|x\| \leqq \sigma$. If the mean value of f with respect to t is zero, then there exists a function $w(t,\theta,x,\epsilon)$ which is multiply periodic in θ of vector period ω; for each fixed ϵ, is almost periodic in t uniformly in θ, x, $\theta \in E^m$, $0 \leqq \|x\| \leqq \sigma$; has a first derivative with respect to t, derivatives of any order with respect to θ, x such that, if

$$g(t,\theta,x,\epsilon) = \frac{\partial w}{\partial t} - f(t,\theta,x,\epsilon)$$

then $\|g(t,\theta,x,\epsilon)\|$, $\|\partial g/\partial\theta\|$, $\|\partial g/\partial x\| \to 0$ as $\epsilon \to 0$ uniformly in t, θ, x. Also, ϵw, $\epsilon\partial w/\partial\theta$, $\epsilon\partial w/\partial x \to 0$ as $\epsilon \to 0$ uniformly in t, θ, x.

Proof. By changing the scale of each of the components of the vector θ we can assume that the vector ω is given by $\omega = (2\pi, \ldots, 2\pi)$. For each fixed ϵ, $0 < \epsilon \leqq \epsilon_0$, the function $f(t,\theta,x,\epsilon)$ satisfies the conditions of Lemma 12-1 with the set $\Lambda = T^m \times (0 \leqq \|x\| \leqq \sigma)$ where $\theta = \text{col}\,(\theta_1, \ldots, \theta_m) \in T^m$ if $-\pi \leqq \theta_j \leqq \pi$, $j = 1, 2, \ldots, m$. Therefore, there exists a continuous function $\zeta(\eta) \to 0$ as $\eta \to 0$ and a function $f_\eta(t,\theta,x,\epsilon)$ defined by (12-5) such that

$$\begin{aligned} \|f_\eta(t,\theta,x,\epsilon)\| &\leqq \eta^{-1}\zeta(\eta) \\ \frac{\partial f_\eta\,(t,\theta,x,\epsilon)}{\partial t} - f(t,x) &= -\eta f_\eta(t,x) \end{aligned} \tag{12-8}$$

for $t \in E$, $(\theta,x) \in \Lambda$. For a fixed a, $0 < a < \pi$ and some fixed integer $q \geqq 1$, consider the two functions $\Delta_a(x)$, $\delta_a(\theta)$ defined by the relations

$$\begin{aligned} \Delta_a(x) &= A_a(1 - a^{-2}\|x\|^2)^{2q} \quad \text{for } \|x\| \leqq a, \ = 0 \text{ for } \|x\| > a \\ \delta_a(\theta) &= \Phi_a(1 - a^{-2}\|\theta\|^2)^{2q} \quad \text{for } \|\theta\| \leqq a, \ = 0 \text{ for } \|\theta\| > a \end{aligned}$$

$-\pi \leqq \theta_j \leqq \pi, j = 1, 2, \ldots, m$ and the definition of $\delta_a(\theta)$ extended to all of E^m by requiring that $\delta_a(\theta)$ is multiply periodic in θ of vector period ω. The constants A_a, Φ_a are determined in such a way that

$$\int_{0\leqq\|x\|\leqq\sigma} \Delta_a(x)\,dx = 1 \qquad \int_{-a}^{a} \delta_a(\theta)\,d\theta = 1$$

* A function $f(\varphi)$, f an m vector, φ an n vector, is said to be multiply periodic in φ with vector period $\omega = \text{col}\,(\omega_1, \ldots, \omega_n)$, $\omega_j > 0, j = 1, 2, \ldots, n$, if the function f is periodic in the jth coordinate of φ with period ω_j.

where the last integral signifies that each component of θ ranges from $-a$ to a. Now consider the function

$$w(t,\theta,x,\epsilon) = \int_{T^m} \int_{0 \leqq \|x'\| \leqq \sigma} \delta_a(\theta - \theta')\, \Delta_a(x - x') f_\eta(t,\theta',x',\epsilon)\, dx'\, d\theta' \quad (12\text{-}9)$$

It is clear that w is multiply periodic in θ of vector period ω.

Notice that the function $\delta_a(\theta - \theta')\, \Delta_a(x - x')$ possesses partial derivatives of order $2q - 1$ with respect to θ, x. Moreover, this function and its derivatives of order $2q - 1$ are bounded in norm by $G(a)/$(area of integration), where $G(a)$ may approach $+\infty$ as $a \to 0$. Since $q \geqq 1$ and the conditions of the lemma are satisfied, it is clear that the partial derivatives of w of any order $\leqq 2q - 1$ with respect to θ, x are bounded by $G(a)\zeta(\eta)\eta^{-1}$.

Up to this point, a, η have been arbitrary, but now we choose them as functions of ϵ in such a way that

$$a_\epsilon \to 0 \qquad \eta_\epsilon \to 0 \qquad \epsilon G(a_\epsilon)\zeta(\eta_\epsilon)\eta_\epsilon^{-1} \to 0 \qquad G(a_\epsilon)\zeta(\eta_\epsilon) \to 0 \quad \text{as } \epsilon \to 0 \quad (12\text{-}10)$$

Since $\epsilon G(a_\epsilon)\zeta(\eta_\epsilon)\eta_\epsilon^{-1} \to 0$ as $\epsilon \to 0$, this proves the last conclusion of the lemma from the remark above. Also, suppose $\rho_0 < \sigma$ and take ϵ^* sufficiently small so that $0 < \epsilon < \epsilon^*$ implies $\rho + a_\epsilon \leqq \sigma$. Then, from the definition of $\Delta_a(x)$, it follows that $\int_{0 \leqq \|x\| \leqq \sigma} \Delta_{a_\epsilon}(x - x')\, dx' = 1$ for every x, $0 \leqq \|x\| \leqq \rho_0$, $0 < \epsilon < \epsilon^*$. Therefore, from relation (12-9) and (12-8) it follows that the function $g(t,\theta,x,\epsilon)$ defined in the statement of the lemma satisfies the relation

$$\begin{aligned} g(t,\theta,x,\epsilon) + \eta_\epsilon w = \int_{T^m} \int_{0 \leqq \|x'\| \leqq \sigma} & \delta_{a_\epsilon}(\theta - \theta')\, \Delta_{a_\epsilon}(x - x') \\ & \cdot [f(t,\theta',x',\epsilon) - f(t,\theta,x,\epsilon)]\, dx'\, d\theta \end{aligned} \quad (12\text{-}11)$$

since

$$\begin{aligned} \int_{T^m} \int_{0 \leqq \|x\| \leqq \sigma} & \delta_{a_\epsilon}(\theta - \theta')\, \Delta_{a_\epsilon}(x - x')\, dx'\, d\theta' \\ & = \int_{T^m} \delta_{a_\epsilon}(\theta - \theta')\, d\theta' = \int_{T^m} \delta_{a_\epsilon}(\theta')\, d\theta' = 1 \end{aligned}$$

On the other hand, since f has first derivatives with respect to θ, x, there exists a function $\lambda(\epsilon)$ bounded for $0 < \epsilon < \epsilon^*$ such that $f(t,\theta,x,\epsilon)$ is Lipschitzian in θ, x with Lipschitz constant $\lambda(\epsilon)$. Since $\delta_a(\theta)$ is periodic of period 2π in each component of θ, we can perform the integration in (12-11) with respect to θ' in any cube of side 2π; in particular, in the cube $-\pi \leqq \theta_j \leqq \pi$, $j = 1, 2, \ldots, m$. Therefore, in $E \times \Lambda$,

$$\|g(t,\theta,x,\epsilon) + \eta_\epsilon w\| \leqq 2\lambda(\epsilon)a_\epsilon \tag{12-12}$$

Integrating by parts in (12-11), it follows that

$$\frac{\partial}{\partial\theta}[g(t,\theta,x,\epsilon) + \eta_\epsilon w] = \int_{T^m}\int_{0\leqq\|x'\|\leqq\sigma} \delta_{a_\epsilon}(\theta - \theta')\,\Delta_{a_\epsilon}(x - x') \cdot \left[\frac{\partial f\,(t,\theta',x',\epsilon)}{\partial\theta} - \frac{\partial f\,(t,\theta,x,\epsilon)}{\partial\theta}\right] dx'\,d\theta'$$

Since $\partial f/\partial\theta$ is assumed to be continuous, there exists a monotone function $\mu(\epsilon) \to 0$ as $\epsilon \to 0$ such that

$$\left\|\frac{\partial(g + \eta_\epsilon w)}{\partial\theta}\right\| \leqq \int_{T^m}\int_{0\leqq\|x'\|\leqq\sigma} \delta_{a_\epsilon}(\theta - \theta')\,\Delta_{a_\epsilon}(x - x')\mu(\epsilon)(\|\theta - \theta'\| + \|x - x'\|)\,dx'\,d\theta' \leqq 2\mu(\epsilon)a_\epsilon \tag{12-13}$$

in $E \times \Lambda$. In the same way, one shows that the partial derivative of $g + \eta_\epsilon w$ with respect to x is bounded by $2\mu(\epsilon)a_\epsilon$. Using the fact that the function $\eta_\epsilon w$ and its partial derivatives with respect to θ, x are bounded by $G(a_\epsilon)\zeta(a_\epsilon)$ in $E \times \Lambda$, it follows from (12-10) to (12-13) that the function w satisfies all the conditions of the lemma except the almost periodicity in t.

To complete the proof of the lemma, it is sufficient, from Bochner's criterion, to show that an arbitrary sequence of real numbers $\{\tau_m\}$ is such that, for any ϵ, $0 < \epsilon \leqq \epsilon_0$, the sequence $\{w(t + \tau_n, \theta, x, \epsilon)\}$ has a subsequence $\{w(t + \tau_{n_k}, \theta, x, \epsilon)\}$ which converges uniformly with respect to $t \in E$, $\theta \in E^m$, $0 \leqq \|x\| \leqq \sigma$. From the assumption on f we can assume that the sequence $\{\tau_n\}$ is such that, for any ϵ, $0 < \epsilon \leqq \epsilon_0$ and any $\delta > 0$, there exists a $K = K(\delta,\epsilon)$ such that

$$\|f(t + \tau_j, \theta, x, \epsilon) - f(t + \tau_k, \theta, x, \epsilon)\| \leqq \delta$$

for $j \geqq k \geqq K$, $t \in E$, $\theta \in E^m$, $0 \leqq \|x\| \leqq \sigma$.

For every $\eta > 0$, $0 < \epsilon \leqq \epsilon_0$, f_η in (12-5) is almost periodic in t uniformly in θ, x. In fact, from (12-5),

$$\|f_\eta(t + \tau_j, \theta, x, \epsilon) - f_\eta(t + \tau_k, \theta, x, \epsilon)\| \leqq \delta\eta^{-1}$$

for $j \geqq k \geqq K$, $t \in E$, $\theta \in E^m$, $0 \leqq \|x\| \leqq \sigma$.

From (12-9),

$$\|w(t + \tau_j, \theta, x, \epsilon) - w(t + \tau_k, \theta, x, \epsilon)\| \leqq \delta G(a)\eta^{-1}$$

for $j \geqq k \geqq K$, $t \in E$, $\theta \in E^m$, $0 \leqq \|x\| \leqq \sigma$, where $G(a)$ is the function defined above. Now let a_ϵ, η_ϵ, $\delta_\epsilon \to 0$ as $\epsilon \to 0$ be chosen so that $\delta_\epsilon G(a_\epsilon)\eta_\epsilon^{-1} = \delta_1 > 0$, independent of ϵ, and the conditions (12-10) are also satisfied. The proof of the lemma is now complete.

Remark 12-1. Let $f(t,\varphi,\theta,x,\epsilon)$, f a k vector, t a scalar, φ an l vector, θ an m vector, x an n vector, ϵ a scalar, be a continuous function of t, φ, θ, x, ϵ for $t \in E$, $\varphi \in E^l$, $\theta \in E^m$, $0 \leqq \|x\| \leqq \sigma$, $0 \leqq \epsilon \leqq \epsilon_0$ and satisfy the conditions of Lemma 12-2 with respect to t, θ, x, ϵ. If f is multiply periodic in φ with vector period $\gamma = (\gamma_1, \ldots, \gamma_l)$ and $\varphi + \tau$ designates the vector, col $(\varphi_1 + \tau, \ldots, \varphi_l + \tau)$, τ a scalar, then define $f_0(t,\varphi,\theta,x,\epsilon)$ to be the mean value of $f(t + \tau, \varphi + \tau, \theta, x, \epsilon)$ with respect to τ; i.e.,

$$f_0(t,\varphi,\theta,x,\epsilon) = \lim_{T \to \infty} T^{-1} \int_0^T f(t + \tau, \varphi + \tau, \theta, x, \epsilon)\, d\tau \qquad (12\text{-}14)$$

This mean value always exists since, for each ϵ, $f(t + \tau, \varphi + \tau, \theta, x, \epsilon)$ is almost periodic in τ uniformly with respect to $t \in E$, $\varphi \in E^l$, $\theta \in E^n$, $0 \leqq \|x\| \leqq \sigma$. If, in addition to the above hypotheses, $f_0(t,\varphi,\theta,x,\epsilon) = 0$ and f has continuous first derivatives with respect to t, φ, then, as shown by Hale [1], there is a function $w(t,\varphi,x,\epsilon)$ which satisfies the conditions of Lemma 12-2 with respect to θ, x, ϵ, is almost periodic in t uniformly in φ, θ, x, and is multiply periodic in φ of vector period γ. Furthermore, the function

$$g(t,\varphi,\theta,x,\epsilon) = \frac{\partial w}{\partial t} + \sum_{j=1}^{l} \frac{\partial w}{\partial \varphi_j} - f(t,\varphi,\theta,x,\epsilon) \qquad (12\text{-}15)$$

and its partial derivatives of first order with respect to θ, x approach zero uniformly as $\epsilon \to 0$. The proof of this fact is modeled after the one above.

In concluding this section, let us discuss the general concept of mean value given by (12-14). If the vector φ does not appear in the function f, then this is the mean value of f with respect to t and then, as remarked previously, f_0 is independent of t. On the other hand, if φ actually appears in the function f, then f_0 may depend upon t, φ. In particular, if $f(t,\varphi,\theta,x,\epsilon)$ is periodic in t with period γ_0 and the frequencies $2\pi/\gamma_0$, $\ldots$, $2\pi/\gamma_l$ are linearly independent over the integers, then $f_0(t,\varphi,\theta,x,\epsilon)$ is independent of t, φ. Of course, these frequencies may be dependent and one still has f_0 independent of t, φ for some functions f.

Suppose φ is a scalar and $f(t,\varphi,\theta,x,\epsilon)$ is multiply periodic in t, φ of periods $2\pi/\mu$, $2\pi/\omega$, respectively. Then f has a Fourier expansion

$$f \sim \Sigma a_{kl} e^{i(k\mu t + l\omega\varphi)}$$

The mean value of $f(t + \tau, \varphi + \tau, \theta, x, \epsilon)$ with respect to τ does not in general coincide with a_{00} but is equal to the sum of those terms a_{kl} for which $k\mu + l\omega = 0$.

The above mean value has been used in the application to differential equations by Krylov and Bogoliubov [1], Bogoliubov and Mitropolski [1], Diliberto [1], and Hale [1].

Lemma 12-2 and the remark above show that it is possible to find a function $w(t,\varphi,\theta,x,y,\epsilon)$ such that one can almost solve the partial differential equation

$$\frac{\partial w}{\partial t} + \sum_j \frac{\partial w}{\partial \varphi_j} - f = 0 \tag{12-16}$$

in the sense that one can make the function on the left-hand side as small as desired. In general, the function w is unbounded as $\epsilon \to 0$. The reason for this is easily understood if one considers a function $f(t)$ which is almost periodic in t. Then the solution of (12-16) would be $a_0 + \int_0^t f(\tau)\,d\tau$, but this integral need not be bounded. If it is bounded, then $w(t)$ of Lemma 12-2 could be chosen as

$$w = a_0 + \int_0^t f(\tau)\,d\tau$$

with a_0 constant.

13: Almost Periodic Solutions—Noncritical Case

The purpose of the present chapter is to state and prove some basic results for the existence of almost periodic solutions for perturbed almost periodic differential systems.

The discussion of almost periodic differential systems is somewhat more complicated than that of periodic systems. Consequently, we begin by discussing the question of the existence of almost periodic solutions of the special system

$$\dot{y} = Ay + q(t,y,\epsilon) \tag{13-1}$$

where ϵ is a parameter, y, q are n vectors, A is an $n \times n$ constant matrix, q is almost periodic in t uniformly in y, ϵ for $\|y\| \leqq R$, $R > 0$, $0 \leqq \epsilon \leqq \epsilon_0$, $\epsilon_0 > 0$, and $q(t,y,\epsilon)$ is Lipschitzian with respect to y for $-\infty < t < \infty$, $0 \leqq \|y\| \leqq R$, $0 \leqq \epsilon \leqq \epsilon_0$.

If we wish to obtain almost periodic solutions of (13-1) by successive approximations, our previous experience has shown that it is necessary to know the behavior of the nonhomogeneous linear system

$$\dot{y} = Ay + f(t) \tag{13-2}$$

for an arbitrary almost periodic function f. If (13-2) is to have an almost periodic solution for an arbitrary f, then certainly the homogeneous system

$$\dot{y} = Ay \tag{13-3}$$

must have no periodic solution (except $y = 0$) of any period; that is, the eigenvalues of the matrix A must all have nonzero real parts. In fact, if A has an eigenvalue with zero real part, then an investigation of the Jordan canonical form of A shows that there exists a periodic function f such that all solutions of (13-2) are unbounded. If all the eigenvalues of A have nonzero real part, we say system (13-3) is *noncritical.* Otherwise, system (13-3) is *critical.*

Lemma 13-1. If system (13-3) is noncritical, then, for every almost periodic function f, there is a unique solution y^* of (13-2) such that y^* is almost periodic and has the same basic frequencies as f. Furthermore, there is a constant K, independent of f, such that

$$\|y^*(t)\| \leqq K \sup_{-\infty<t<\infty} \|f(t)\| \qquad -\infty < t < \infty$$

and the function $y^*(t)$ is given by formula (13-6) below.

Proof. Without loss in generality, we can assume that

$$A = \operatorname{diag}(A_+, A_-)$$

where A_+ has all eigenvalues with positive real parts and A_- has all eigenvalues with negative real parts. Define the $n \times n$ matrix $J_A(t)$ by the relation

$$\begin{aligned} J_A(t) &= -\begin{bmatrix} e^{-A_+t} & 0 \\ 0 & 0 \end{bmatrix} \qquad \text{for } t > 0 \\ J_A(t) &= \begin{bmatrix} 0 & 0 \\ 0 & e^{-A_-t} \end{bmatrix} \qquad \text{for } t < 0 \\ J_A(-0) &- J_A(+0) = I \end{aligned} \tag{13-4}$$

It is now clear that there exist positive constants α, β such that

$$|J_A(t)| \leqq \beta e^{-\alpha|t|} \qquad -\infty < t < \infty \tag{13-5}$$

With this definition of $J_A(t)$, define the function $y^*(t)$ by

$$y^*(t) = \int_{-\infty}^{\infty} J_A(\tau) f(t+\tau)\, d\tau = \int_{-\infty}^{\infty} J_A(s-t) f(s)\, ds \tag{13-6}$$

The integral on the right-hand side of (13-6) is meaningful since $J_A(t)$ satisfies (13-5) and $f(t)$ is bounded. In fact,

$$\|y^*(t)\| \leqq \sup_{-\infty<t<\infty} \|f(t)\| \int_{-\infty}^{\infty} \beta e^{-\alpha|\tau|}\, d\tau = \frac{2\beta}{\alpha} \sup_{-\infty<t<\infty} \|f(t)\|$$

By direct computation, one shows that $y^*(t)$ is a solution of (13-2) and, in fact, is the only bounded solution of (13-2) in the interval $-\infty < t < \infty$.

Now suppose there exists a sequence of real numbers t_m, $m = 1, 2, \ldots$, such that

$$\|f(t + t_m) - f(t)\| \to 0 \text{ as } m \to \infty$$

uniformly in t; that is, for any $\epsilon > 0$, there exists an integer $M = M(\epsilon)$ such that

$$\|f(t + t_m) - f(t)\| < \epsilon \qquad m \geqq M \qquad -\infty < t < \infty$$

From (13-6),

$$\|y^*(t + \tau_m) - y^*(t)\| \leqq \frac{2\beta\epsilon}{\alpha} \qquad m \geqq M \qquad -\infty < t < \infty$$

Since f is almost periodic, it follows from the remarks about basic frequencies in Chap. 12 that y^* is almost periodic and the lemma is proved.

The trick of defining $y^*(t)$ as in (13-6) is essentially due to Liapunov [1].

Theorem 13-1. Suppose (13-3) is noncritical, $q(t,y,\epsilon)$ is almost periodic in t uniformly in y, $0 \leqq \|y\| \leqq R$, for each fixed ϵ, $0 < \epsilon \leqq \epsilon_0$, and there exists a function $\eta(\epsilon,\rho)$, $M(\epsilon)$, continuous and nondecreasing in ϵ, ρ for $0 \leqq \epsilon \leqq \epsilon_0$, $0 \leqq \rho \leqq R$, such that $\eta(0,0) = 0$, $M(0) = 0$, and

$$\begin{aligned} \|q(t,y_1,\epsilon) - q(t,y_2,\epsilon)\| &\leqq \eta(\epsilon,\rho)\|y_1 - y_2\| \\ \|q(t,0,\epsilon)\| &\leqq M(\epsilon) \end{aligned} \tag{13-7}$$

for $-\infty < t < \infty$, $\|y_1\| \leqq \rho$, $\|y_2\| \leqq \rho$, $0 \leqq \epsilon \leqq \epsilon_0$. Under these conditions, there exist constants $\sigma > 0$, $\epsilon_1 > 0$ such that there is a solution $y^*(t,\epsilon), 0 < \epsilon \leqq \epsilon_1$ of (13-1) with y^* almost periodic in t for each fixed ϵ in $0 < \epsilon \leqq \epsilon_1$, $y^*(t,\epsilon)$ continuous in ϵ for $0 < \epsilon \leqq \epsilon_1$, $y^*(t,\epsilon) \to 0$ as $\epsilon \to 0$ uniformly in t, $-\infty < t < \infty$, and y^* is the only solution of (13-1) which remains in the region $0 \leqq \|y\| \leqq \sigma$ for $-\infty < t < \infty$. If all the eigenvalues of the matrix A have negative real parts, then $y^*(t,\epsilon)$ is asymptotically stable in the sense of Liapunov. If one eigenvalue has a positive real part, then $y^*(t,\epsilon)$ is unstable.

Proof. Let $y^{(0)} = 0$, and define recursively $y^{(k+1)}(t)$, $k = 0, 1, 2, \ldots$, to be the unique almost periodic solution of

$$\dot{z} = Az + q(t,y^{(k)}(t),\epsilon)$$

which is guaranteed by Lemma 13-1. From the estimate of $y^{(k+1)}$ given in Lemma 13-1 and relation (13-7), it is easy to see that the sequence $\{y^{(k)}(t)\}$ converges uniformly to a function $y^*(t,\epsilon)$ which is continuous in ϵ for $0 < \epsilon \leqq \epsilon_1$, $\epsilon_1 > 0$, y^* is almost periodic in t for each fixed ϵ in $0 < \epsilon \leqq \epsilon_1$, and $y^*(t,\epsilon) \to 0$ as $\epsilon \to 0$ uniformly in t.

To prove uniqueness of this almost periodic solution in a neighborhood of $y = 0$, one can use the stability properties of the almost periodic solution and the fact that $y^*(t,\epsilon) \to 0$ as $\epsilon \to 0$ uniformly in t. The stability properties can be obtained by using the variational equations of $y^*(t,\epsilon)$ and the well-known theorem of stability of Liapunov (see Coddington and Levinson [1, Chap. 13]). Rather than apply these known results directly, we indicate another method of investigating these stability properties. This method does not depend on the variational equations and is applicable to the more general situation of Chap. 15. The complete details of this proof may be found in Bogoliubov and Mitropolski [1, pp. 371*ff*.].

With $J_A(t)$ defined as in (13-4), consider the integral equation

$$y(t) = \int_{t_0}^{\infty} J_A(\tau - t)q(\tau,y(\tau),\epsilon)\, d\tau + J_A(t_0 - t)b \qquad t \geqq t_0 \geqq 0 \quad (13\text{-}8)$$

where $t_0 \geqq 0$ is given and b is a given constant vector in E^n.

For any $\sigma_1 < R$, there exists an $\epsilon_2 \leqq \epsilon_1$, $\sigma_2 \leqq \sigma_1$ such that, for $0 \leqq \epsilon \leqq \epsilon_2$ and any given b, $\|b\| \leqq \sigma_2$, this integral equation has a unique solution $y(t,t_0,b,\epsilon)$ such that $\|y(t,t_0,b,\epsilon)\| \leqq \sigma_1$, for all $t \geqq t_0$. Furthermore, if $\|b_1\| \leqq \sigma_2$, $\|b_2\| \leqq \sigma_2$, then

$$\|y(t,t_0,b_1,\epsilon) - y(t,t_0,b_2,\epsilon)\| \leqq \mu(\epsilon_2,\sigma_2) \left\{\exp\left[-\frac{\alpha}{2}(t - t_0)\right]\right\} \|b_1 - b_2\|$$

for all $t \geqq t_0$ and $\mu(\epsilon_2,\sigma_2)$ is continuous and $\to 0$ as $\epsilon_2 \to 0$, $\sigma_2 \to 0$.

Furthermore, the solutions of this integral equation satisfy (13-1). Conversely, there exists a $\sigma_1^0 > 0$ such that, if $y(t,t_0,y_0,\epsilon)$

$$y(t_0,t_0,y_0,\epsilon) = y_0$$

is a solution of (13-1) with $\|y_0\| \leqq \sigma_2$, $\|y(t,t_0,y_0,\epsilon)\| \leqq \sigma_1$, $t \geqq t_0$, for all $\sigma_1 \leqq \sigma_1^0$, then $y(t,t_0,y_0,\epsilon)$ satisfies the above integral equation for some b.

But, for any $\sigma_1 > 0$ and sufficiently small, our almost periodic solution $y^*(t,\epsilon)$ of (13-1) satisfies $\|y^*(t,\epsilon)\| \leqq \sigma_1$ for all t. Consequently, there exists a b^* such that $y^*(t,\epsilon) = y(t,t_0,b^*,\epsilon)$, a solution of the integral equation. This implies that

$$\|y(t,t_0,b,\epsilon) - y^*(t,\epsilon)\| \leqq \mu(\epsilon_2,\sigma_2) \left\{\exp\left[-\frac{\alpha}{2}(t - t_0)\right]\right\} \|b - b^*\|$$

for all $t \geqq t_0$ and solutions $y(t,t_0,b,\epsilon)$ of the integral equation with $\|b\| \leqq \sigma_2$; or, $y^*(t,\epsilon)$ is exponentially asymptotically stable in this class of functions.

This allows one to conclude that the solution $y^*(t,\epsilon)$ is unique in a neighborhood of $y = 0$.

Theorem 13-2. Suppose system (13-3) is noncritical and let $f(t,y,\epsilon)$, $g(t,y,\epsilon)$ be given n-vector functions which are almost periodic in t uniformly in y, $0 \leqq \|y\| \leqq R$, $R > 0$ for each fixed ϵ, $0 < \epsilon \leqq \epsilon_0$, are continuous in t, y, ϵ and Lipschitzian with respect to y for $-\infty < t < \infty$, $0 \leqq \|y\| \leqq R$, $0 < \epsilon \leqq \epsilon_0$. If $f(t,y,\epsilon)$ satisfies (13-7), the second partials of g with respect to y are continuous and

$$\lim_{T\to\infty} T^{-1} \int_0^T g(\tau,y,\epsilon)\, d\tau = 0 \tag{13-9}$$

then there are $\sigma > 0$, $\epsilon_1 > 0$, $\epsilon_1 \leqq \epsilon_0$, such that the system of equations

$$\dot{y} = \epsilon[Ay + f(t,y,\epsilon) + g(t,y,\epsilon)] \tag{13-10}$$

has a solution $y^*(t,\epsilon)$ which is almost periodic in t for each fixed ϵ, $0 < \epsilon \leqq \epsilon_1$, $y^*(t,\epsilon) \to 0$ as $\epsilon \to 0$ uniformly in t and, for $0 < \epsilon \leqq \epsilon_1$, this solution is the only solution which remains in $0 \leqq \|y\| \leqq \sigma$ for $-\infty < t < \infty$. If all the eigenvalues of A have negative real parts, then y^* is asymptotically stable and if one eigenvalue has a positive real part, then y^* is unstable.

Proof. From (13-9) and Lemma 12-2, for any $\epsilon > 0$, there exists a function $w(t,y,\epsilon)$ such that

$$\left\| \frac{\partial w\,(t,y,\epsilon)}{\partial t} - g(t,y,\epsilon) \right\| \leqq \sigma(\epsilon)$$

for all t, y and $\sigma(\epsilon) \to 0$ as $\epsilon \to 0$. Furthermore, from Lemma 12-2 and (13-7), the transformation

$$y = z + \epsilon w(t,z,\epsilon)$$

is one-to-one and, if applied to (13-10), yields an equation in z of the form

$$\dot{z} = \epsilon[Az + q(t,z,\epsilon)]$$

where q satisfies the condition of Theorem 13-1. If $\tau = \epsilon t$, then the result follows from Theorem 13-1.

If one does not wish to apply Theorem 13-1 to complete the proof of Theorem 13-2, then it can be proved directly as follows: Let $J_A(t)$ be defined as in (13-4) and consider the method of successive approximations

$$y^{(0)} = 0$$

$$y^{(k+1)}(t) = \epsilon \int_{-\infty}^{\infty} J_A(\epsilon\tau) q(t+\tau, y^{(k)}(t+\tau),\epsilon)\, d\tau$$

From the estimate (13-5), it follows that $\|J_A(\epsilon t)\| \leqq \beta e^{-\alpha\epsilon|t|}$, $-\infty < t < \infty$, $0 < \epsilon \leqq \epsilon_1$ and one can easily prove convergence of this method as in the proof of Theorem 13-1.

An equivalent statement of Theorem 13-2 is

Corollary 13-1. If system (13-3) is noncritical and f, g satisfy the conditions of Theorem 13-2, there exist an $\omega_0 > 0$ and a $\sigma > 0$ such that the system of equations

$$\frac{dy}{dt} = Ay + f(\omega t, y, \omega^{-1}) + g(\omega t, y, \omega^{-1}) \tag{13-11}$$

has a solution $y^*(t,\omega)$ which is almost periodic in t for $\omega \geqq \omega_0$, $y^*(t,\omega) \to 0$ as $\omega \to \infty$ and, for every finite $\omega \geqq \omega_0$, this solution is the only solution which remains in $0 \leqq \|y\| \leqq \sigma$ for $-\infty < t < \infty$. If all the eigenvalues of A have negative real parts, then y^* is asymptotically stable and if one eigenvalue has a positive real part, then y^* is unstable.

Proof. If $\omega t = \tau$, $\epsilon = \omega^{-1}$, then system (13-11) is the same as (13-10) of Theorem 13-2.

By using a proof similar to the above proof of Theorem 13-1, Langenhop [1] has proved a special case of Corollary 13-1 when $g(t,y) = h(t)$ where $\int^t h(\tau)\,d\tau$ is bounded. In this case, as remarked in Chap. 12, the function w used in the proof of Theorem 13-2 can be taken as $w = \int^t h(\tau)\,d\tau$. For further results on system (13-11), see Demidovich [1].

It is clear that one could give a single result which includes both Theorem 13-1 and Corollary 13-1 for an equation of the form

$$\dot{y} = Ay + q(t,y,\epsilon) + g(\omega t, y)$$

where q satisfies the conditions (13-7) and g satisfies (13-9) while ϵ is sufficiently small and ω is sufficiently large.

A simple example illustrating the above result is the forced van der Pol equation

$$\begin{aligned} \dot{x}_1 &= x_2 \\ \dot{x}_2 &= -x_1 + k(1 - x_1^2)x_2 + \epsilon g^*(\omega t) \end{aligned} \tag{13-12}$$

where $\omega > 0$, $k \neq 0$ are constants, x_1, x_2 are scalars, and $g^*(\tau)$ is periodic in τ. If $x = \operatorname{col}(x_1, x_2)$ then this system can be written in vector form

$$\dot{x} = Ax + f(t,x) + \epsilon g(\omega t)$$

where $A = \begin{bmatrix} 0 & 1 \\ -1 & k \end{bmatrix} \quad f = \begin{bmatrix} 0 \\ -kx_1^2x_2 \end{bmatrix} \quad g = \begin{bmatrix} 0 \\ g^* \end{bmatrix}$

Since $k \neq 0$, the eigenvalues of A have nonzero real parts. Consequently, from Theorem 13-1, for any given g^* and ϵ sufficiently small, there exists an almost periodic solution of (13-12) which approaches zero as $\epsilon \to 0$. If $\lim_{T\to\infty} T^{-1} \int_0^T g(\tau)\, d\tau = 0$, then by Corollary 13-1, for ω sufficiently large, there exists an almost periodic solution of (13-12) which approaches zero as $\omega \to \infty$. In both cases, this solution is asymptotically stable (unstable) if $k < 0$ ($k > 0$).

Theorem 13-3. Suppose system (13-3) is noncritical and let $f(t,y,\epsilon)$ be a given n-vector function which is almost periodic in t, uniformly with respect to y, $0 \leqq \|y\| \leqq R$, for each fixed ϵ, $0 \leqq \epsilon \leqq \epsilon_1$, and is continuous in t, y, ϵ and Lipschitzian with respect to y for $-\infty < t < \infty$, $0 \leqq \|y\| \leqq R$, $0 \leqq \epsilon \leqq \epsilon_1$. If $f(t,y,\epsilon)$ satisfies relation (13-7), then there exist an $\epsilon_1 > 0$ and a $\sigma > 0$ such that the system

$$\epsilon\dot{y} = Ay + f(t,y,\epsilon) \tag{13-13}$$

has a solution $y^*(t,\epsilon)$, which is almost periodic in t for each fixed ϵ, $0 \leqq \epsilon \leqq \epsilon_1$, is continuous in ϵ for $0 \leqq \epsilon \leqq \epsilon_1$, $y^*(t,0) = 0$, and this solution y^* is the only solution of (13-13) which remains in the region $0 \leqq \|y\| \leqq \sigma$ for $-\infty < t < \infty$. If all the eigenvalues of A have negative real parts, then y^* is asymptotically stable and if one eigenvalue has a positive real part, then y^* is unstable.

Proof. The proof follows along the same lines as the proof of Theorem 13-1. In fact, let $J_A(t)$ be defined by (13-4) and consider the method of successive approximations

$$\begin{aligned} y^{(0)} &= 0 \\ y^{(k+1)}(t) &= \int_{-\infty}^{\infty} J_A\left(\frac{\tau}{\epsilon}\right)\left[\frac{1}{\epsilon}f(t+\tau, y^{(k)}(t+\tau),\epsilon)\right] d\tau \qquad k = 0, 1, 2, \ldots \end{aligned} \tag{13-14}$$

The function $y^{(k+1)}(t)$ is the unique almost periodic solution of the equation

$$\epsilon\dot{y} = Ay + f(t,y^{(k)},\epsilon)$$

From the estimate (13-5), $\|J_A(t/\epsilon)\| \leqq \beta e^{-\alpha|t|/\epsilon}$, $-\infty < t < \infty$, $0 < \epsilon \leqq \epsilon_0$. It is now easy to show from this estimate on J_A that the method of successive approximations converges uniformly to a function $y^*(t,\epsilon)$ for $0 < \epsilon \leqq \epsilon_1$, $0 < \epsilon_1 \leqq \epsilon_0$, ϵ_1 sufficiently small and, furthermore, that $y^*(t,\epsilon)$ is almost periodic in t for each fixed ϵ, $0 <$

$\epsilon \leqq \epsilon_1$. Observing that system (13-13) for $\epsilon = 0$ has the solution $y = 0$ one can define $y^*(t,0) = 0$ and the remainder of the proof is the same as that of Theorem 13-1.

By combining the ideas used in the proofs of Theorems 13-1, 13-2, and 13-3, one can easily prove the following result, which contains all the preceding theorems as special cases.

Consider the system of equations

$$\begin{aligned} \dot{x} &= \epsilon[Ax + X(t,x,y,z,\epsilon) + X_1(t,x,y,z,\epsilon)] \\ \dot{y} &= By + Y(t,x,y,z,\epsilon) \\ \epsilon\dot{z} &= Cz + Z(t,x,y,z,\epsilon) \end{aligned} \tag{13-15}$$

where x, y, z are vectors and the following hypotheses are satisfied:

Hypothesis 1. For each fixed ϵ, all functions are almost periodic in t uniformly with respect to x, y, z for $\|x\| \leqq R$, $\|y\| \leqq R$, $\|z\| \leqq R$, where R is a positive constant.

Hypothesis 2. X, Y, Z are continuous in t, x, y, z, ϵ, Lipschitzian in x, y, z for $-\infty < t < \infty$, $\|x\|$, $\|y\|$, $\|z\| \leqq R$, $0 < \epsilon \leqq \epsilon_1$, and the Lipschitz constant approaches zero as $\|x\|$, $\|y\|$, $\|z\|$, $\epsilon \to 0$. Furthermore, for $x = 0$, $y = 0$, $z = 0$, the functions X, Y, Z are bounded by a continuous function $M(\epsilon)$ with $M(0) = 0$.

Hypothesis 3. The eigenvalues of each of the matrices A, B, C, have nonzero real parts.

Hypothesis 4. $X_1(t,x,y,z,\epsilon)$ is continuous together with its first partial derivatives with respect to x, y, z for $-\infty < t < \infty$, $\|x\|$, $\|y\|$, $\|z\| \leqq R$, $0 < \epsilon \leqq \epsilon_1$ and

$$\lim_{T\to\infty} T^{-1} \int_0^T X_1(\tau,x,y,z,\epsilon)\, d\tau = 0$$

Theorem 13-4. If system (13-15) satisfies hypotheses 1 to 4, then there exist $\epsilon_2 > 0$, $\sigma > 0$, such that, for each ϵ, $0 < \epsilon \leqq \epsilon_2$, system (13-15) has a solution which is almost periodic in t and approaches zero uniformly in t as $\epsilon \to 0$ and is unique in the region $0 \leqq \|x\| + \|y\| + \|z\| \leqq \sigma$. If all the eigenvalues of A, B, C have negative real parts, this almost periodic solution is asymptotically stable. If one eigenvalue has a positive real part, it is unstable.

The conclusions of Theorem 13-4 remain valid in some cases when B, C depend upon t, say, $B = B(t)$, $C = C(t)$. In fact, hypothesis 3 can be replaced by the assumption that the zero solution of the system

$$\dot{y} = B(t)y$$

is uniformly asymptotically stable and that all the eigenvalues $\lambda(t)$ of the matrix $C(t)$ satisfy Re $\lambda(t) \leqq -2\gamma < 0$, $-\infty < t < \infty$, where γ is a constant. This new hypothesis implies that, if $\Phi(t)$, $\Psi(t)$ are fundamental solutions of $\dot{y} = B(t)y$, $\epsilon\dot{z} = C(t)z$, respectively, then there is a constant β such that

$$\|\Phi(t)\| \leqq \beta \exp\,[-\gamma(t - t_0)] \qquad \|\Psi(t)\| \leqq \beta \exp\,[-\gamma(t - t_0)/\epsilon]$$

for ϵ sufficiently small.

For the variable x absent in (13-15), a more general result than the above has been given by Hale and Seifert [1]. See, also, Volosov [1].

14: Periodic Solutions Revisited

In some previous chapters, we have shown how to obtain periodic solutions of a system of equations

$$\dot{x} = Ax + \epsilon f(t,x) \tag{14-1}$$

where ϵ is a parameter, x, f are n vectors, A is a constant $n \times n$ matrix, and $f(t,x)$ is periodic in t of period T. For this part of the discussion we assume $f(t,x)$ is sufficiently smooth in t, x so that all formal operations below are valid. More precise results will be stated later. It was shown in Chap. 5 that, if the system

$$x' = Ax \tag{14-2}$$

has no periodic solution of period T except the solution $x = 0$, then there always is a periodic solution of (14-1) of period T for ϵ sufficiently small and this solution approaches zero as ϵ approaches 0. If (14-2) has some periodic solutions of period T, then it is necessary that the function f in (14-1) satisfy some auxiliary conditions in order to have a periodic solution of (14-1) (see Chap. 9). To motivate in a simpler way the discussion of later sections, let us have a relook at this problem for the simple case where all the solutions of (14-2) are periodic of period T. For example, if $T = 2\pi/\omega$ and all the eigenvalues of A have simple elementary divisors and these eigenvalues are integer multiples of ω, then this will be the case. Under this assumption, the matrix e^{At} is periodic of period T. If we introduce new variables y by the transformation

$$x = e^{At}y \tag{14-3}$$

then the new differential equation for y is

$$\dot{y} = \epsilon e^{-At} f(t, e^{At} y) \overset{\text{def}}{\equiv} \epsilon q(t,y) \tag{14-4}$$

where $q(t,y)$ is periodic in t of period T. The basic problem now is to determine a periodic solution of (14-4) which reduces to a constant as $\epsilon \to 0$, since this will give a periodic solution of (14-1) by the transformation (14-3).

Along with system (14-4) consider the autonomous system (called the "averaged" system)

$$\dot{y} = \epsilon q_0(y) \qquad q_0(y) = \frac{1}{T} \int_0^T q(t,y)\, dt \tag{14-5}$$

For the simple case under consideration, Theorem 6-4 implies that, if there exists a vector y_0 such that

$$q_0(y_0) = 0 \qquad \det\,[\partial q_0\,(y_0)/\partial y] \neq 0 \tag{14-6}$$

then there is a periodic solution of (14-1) which for $\epsilon = 0$ is $e^{At}y_0$. Let us interpret (14-6) in terms of the averaged equations (14-5). The vector y_0 is an equilibrium point of (14-5). The linear variational equation with respect to y_0 is

$$\dot{z} = \epsilon \frac{\partial q_0\,(y_0)}{\partial y}\, z \tag{14-7}$$

The second hypothesis in (14-6) implies that, for $\epsilon \neq 0$ and sufficiently small, there is no periodic solution of (14-7) of period T except $z = 0$. Consequently, we can restate conditions (14-6) in terms of the averaged equations as follows: If there exists an equilibrium point y_0 of (14-5) such that, for $\epsilon \neq 0$ and sufficiently small, the linear variational equation has no periodic solution of period T except the trivial one, then there is a periodic solution of (14-4) which for $\epsilon = 0$ coincides with y_0.

We briefly sketch a proof of this fact. Let $u(t,y)$ be a periodic function of period T such that

$$\frac{\partial u\,(t,y)}{\partial t} = q(t,y) - q_0(y) \tag{14-8}$$

Such a function exists since the right-hand side of (14-8) is a periodic function of period T of mean value zero. For ϵ sufficiently small, the transformation

$$y = z + \epsilon u(t,z) \tag{14-9}$$

is one-to-one. If y in (14-4) is replaced by (14-9), then

$$\left[I + \epsilon \frac{\partial u\,(t,z)}{\partial y}\right] \dot{z} = \epsilon q_0(z + \epsilon u(t,z)) + \epsilon \left[q(t,z) - q_0(z) - \frac{\partial u\,(t,z)}{\partial t}\right] + \epsilon[q(t,z + \epsilon u(t,z)) - q(t,z)]$$

From (14-8), it follows that

$$\dot{z} = \epsilon q_0(z) + \epsilon^2 q^*(t,z,\epsilon) \tag{14-10}$$

where $q^*(t,z,\epsilon)$ is periodic in t of period T. What we have shown is that, if we consider equation (14-4) as a perturbation of (14-5), namely, as

$$\dot{y} = \epsilon q_0(y) + \epsilon[q(t,y) - q_0(y)]$$

then the transformation (14-9) has the effect of making the perturbation term of order ϵ^2. To prove the existence of a periodic solution of (14-10) under the hypotheses (14-6) is now very easy and will not be supplied (see Theorem 5-2).

Now let us weaken the hypotheses on the matrix A; namely, let us suppose only that all the eigenvalues of A are purely imaginary and have simple elementary divisors. If A is a real matrix and $\pm i\omega_j$ are the eigenvalues of A, then the solutions of (14-2) are of the form

$$\Sigma a_j \cos \omega_j t + b_j \sin \omega_j t$$

for some constants a_j, b_j. These functions may not be periodic since ω_j/ω_k may be irrational for some j, k. Functions of this type are called almost periodic (see Definition 12-1). Consequently, if we make the transformation (14-3) the system of equations (14-4) becomes almost periodic in t. For almost periodic functions $q(t,y)$ of this type, it is known, as remarked before, that the following limit exists uniformly with respect to t and is independent of t:

$$q_0(y) = \lim_{T\to\infty} \frac{1}{T} \int_t^{t+T} q(\tau,y)\,d\tau \tag{14-11}$$

Finally, if we consider the "averaged" equations

$$\dot{y} = \epsilon q_0(y) \tag{14-12}$$

with $q_0(y)$ defined in (14-11), then we can ask whether or not it is possible to determine almost periodic solutions of (14-4) which degenerate to equilibrium points of (14-12) for $\epsilon = 0$. Now if one could make a transformation of the form (14-9) which would have the effect of making system (14-4) a perturbation of system (14-12), then the results of Chap. 13 would apply. But the essential property of the transformation (14-9) was that

$$\frac{\partial u}{\partial t} = g(t,z) - g_0(z)$$

As remarked in Chap. 12, for almost periodic functions one cannot solve this equation for a bounded u, but by Lemma 12-2 one can make the difference $\partial u/\partial t - [g(t,z) - g_0(z)]$ as small as desired with a function u such that ϵu is bounded and actually approaches zero as $\epsilon \to 0$. Consequently, the transformation

$$y = z + \epsilon u(t,z,\epsilon)$$

with u given by Lemma 12-2 will have the properties mentioned to solve our problem and Chap. 13 applies. We do not delve into this particular problem any further at this time but choose to discuss a very general method of averaging in the next chapter and return to particular cases and examples later.

15: Integral Manifolds—Averaging

The present chapter is devoted to a theoretical discussion of integral manifolds and a method of averaging. Only sketches of the proofs of the results will be given since they are rather complicated and are easily accessible in the literature. Applications will follow in later chapters.

First of all, we give an analytic definition of an integral manifold of a system of differential equations

$$\dot{x} = X(t,x) \tag{15-1}$$

where x, X are n vectors, $X(t,x)$ is continuous in t, x for $-\infty < t < \infty$, $x \in U$, an open set in E^n.

In the (x,t) space, suppose there exists a surface S of dimension $s + 1$ which may be described parametrically by means of the equations

$$S = \{(x,t)|x = f(t, C_1, \ldots, C_s), -\infty < t < \infty\} \tag{15-2}$$

where $s \leqq n$, $f(t, C_1, \ldots, C_s)$ is a continuous function of $t, C_1, \ldots, C_s$ in the whole range of their variation.

The surface S will be called an $(s + 1)$-*dimensional integral manifold* of system (15-1) if any solution of (15-1), $x(t,t_0,x_0)$, $x(t_0,t_0,x_0) = x_0$, with $(x_0,t_0) \in S$ has the property that $(x(t,t_0,x_0),t) \in S$ for all t, $-\infty < t < \infty$.

The simplest type of integral manifold of system (15-1) would be the set S consisting of those points x, t for which $x(t,t_0,x_0)$, $x(t_0,t_0,x_0) = x_0$, is a solution of (15-1), which is defined for $-\infty < t < \infty$. Another more interesting one is the following: Suppose $X(t,x)$ in (15-1) is independent of t; that is, consider the system

$$\dot{x} = X(x) \tag{15-3}$$

and suppose that this equation has a nonconstant periodic solution $x = x^0(t)$ of period 2π. Since (15-3) is autonomous, the function $x = x^0(t + \varphi)$ is also a periodic solution of (15-3) for every arbitrary constant φ. Furthermore, for any φ, the pair $(x^0(t + \varphi),t)$ lies on the cylinder S in $(n + 1)$-dimensional (x,t) space defined parametrically by the equation

$$S = \{(x,t)|x = x^0(\theta), 0 \leqq \theta \leqq 2\pi, -\infty < t < \infty\} \tag{15-4}$$

Furthermore, any solution of (15-3), $x(t, t_0,x_0)$, $x(t_0,t_0,x_0) = x_0$, with $(x_0,t_0) \in S$ must coincide with one of the periodic motions above and the cylinder S is an integral manifold of (15-3) of dimension 2.

Now, consider the perturbed system

$$\dot{x} = X(x) + \epsilon X^*(t,x) \tag{15-5}$$

where for $\epsilon = 0$ the system has a periodic solution $x^0(t)$ of period 2π and the perturbation term $X^*(t,x)$ is a bounded function for $-\infty < t < \infty, x \in U$. Under what conditions on the function $X(x)$ will the solutions of (15-3) and (15-5) be "essentially" the same for ϵ sufficiently small? Of course, one cannot begin to answer such a question without first clarifying the word "essentially." Suppose, for example, that the periodic solution $x^0(t)$ of (15-3) is exponentially asymptotically orbitally stable; that is, the cylinder S in (15-4) is exponentially asymptotically stable. Since the cylinder S is filled with a one-parameter family of periodic solutions differing only by a shift in phase, one could not hope that, under small general perturbations $\epsilon X^*(t,x)$, each particular periodic solution on S enjoys a property of stability. However, it seems reasonable to suppose that there is another integral manifold S_ϵ of (15-5) which is stable and $S_\epsilon \to S$ as $\epsilon \to 0$. It is this type of problem which we wish to discuss.

The specific problem mentioned above for system (15-5) will be discussed further in Chap. 16. As we shall see, this question, together with many others, can be reduced to a discussion of systems of differential equations of the form

$$\begin{aligned}\dot{\theta} &= d(\epsilon) + \Theta(t,\theta,x,y,\epsilon)\\ \dot{x} &= \epsilon Cx + \epsilon X(t,\theta,x,y,\epsilon)\\ \dot{y} &= Ay + Y(t,\theta,x,y,\epsilon)\end{aligned} \tag{15-6}$$

where ϵ is a real parameter, $d(\epsilon)$ is a constant vector continuous in ϵ for $0 < \epsilon \leqq \epsilon_0$, θ, x, y are k, m, n vectors, respectively. For any σ, μ define the set $\Sigma_{\sigma,\mu}$ as follows:

$$\Sigma_{\sigma,\mu} = \{(t,\theta,x,y)|t \in E, \theta \in E^k, 0 \leqq \|x\| \leqq \sigma, 0 \leqq \|y\| \leqq \mu\} \tag{15-7}$$

The following hypotheses will also be needed:

There exist positive constants ρ_1, ρ_2 such that the functions Θ, X, Y are continuous on $\Sigma_{\rho_1,\rho_2} \times (0,\epsilon_0]$ and are multiply periodic in θ with vector period ω; that is, if $\theta = \text{col}\,(\theta_1, \ldots, \theta_k)$, $\omega = (\omega_1, \ldots, \omega_k)$, $\omega_j > 0$, then Θ, X, Y are periodic in θ_j with period ω_j, $j = 1,2, \ldots, k$. (15-8)

There exists a function $M(\epsilon)$ continuous in ϵ for $0 < \epsilon \leqq \epsilon_0$ such that $M(\epsilon) \to 0$ as $\epsilon \to 0$ and each of the functions Θ, X, Y is bounded by $M(\epsilon)$ on $\Sigma_{0,0}$. (15-9)

There are functions $\eta(\epsilon,\sigma,\mu)$, $\gamma(\epsilon,\sigma,\mu)$, continuous in ϵ, σ, μ for $0 < \epsilon \leqq \epsilon_0$, $0 \leqq \sigma \leqq \rho_1$, $0 \leqq \mu \leqq \rho_2$, $\eta(\epsilon,\sigma,\mu) \to 0$ as $\epsilon, \sigma, \mu \to 0$, $\eta(\epsilon,0,0) = o(\epsilon)$, $\gamma(\epsilon,o,o) = o(\epsilon)$ as $\epsilon \to 0$ such that the function Θ is Lipschitzian in θ, x, y on $\Sigma_{\sigma,\mu} \times (0,\epsilon_0]$ with Lipschitz constant $\eta(\epsilon,\sigma,\mu)$ in θ and $\gamma(\epsilon,\sigma,\mu)$ in x, y. If the vector x is absent in (15-6) then the condition $\eta(\epsilon,0,0) = o(\epsilon)$, $\gamma(\epsilon,o,o) = o(\epsilon)$ as $\epsilon \to 0$ is unnecessary. (15-10)

There are functions $\lambda(\epsilon,\sigma,\mu)$, $\delta(\epsilon,\sigma,\mu)$ continuous in ϵ, σ, μ for $0 < \epsilon \leqq \epsilon_0$, $0 \leqq \sigma \leqq \rho_1$, $0 \leqq \mu \leqq \rho_2$, $\lambda(\epsilon,\sigma,\mu) \to 0$, $\delta(\epsilon,\sigma,\mu) \to o$, $\lambda(o,\sigma,\mu) = o(\sigma + \mu)$ as $\epsilon,\sigma,\mu \to 0$ such that the functions X, Y are Lipschitzian in θ, x, y on $\Sigma_{\sigma,\mu} \times (0,\epsilon_0]$ with Lipschitz constant $\lambda(\epsilon,\sigma,\mu)$ in θ and $\delta(\epsilon,\sigma,\mu)$ in x, y. (15-11)

The eigenvalues of the constant matrices A, C have nonzero real parts. (15-12)

Theorem 15-1. If system (15-6) satisfies conditions (15-8) to (15-12), then there exist an $\epsilon_1 > 0$, scalar functions $D(\epsilon)$, $\Delta(\epsilon)$, and vector functions $f(t,\theta,\epsilon)$, $g(t,\theta,\epsilon)$, of dimensions m, n, respectively, which are continuous in t, θ, ϵ for $t \in E$, $\theta \in E^k$, $0 < \epsilon \leqq \epsilon_1$; $D(\epsilon) \to 0$, $\Delta(\epsilon) \to 0$ as $\epsilon \to 0$; f, g multiply periodic in θ of vector period ω;

$$\|f(t,\theta,\epsilon)\| \leqq D(\epsilon) \qquad \|g(t,\theta,\epsilon)\| \leqq D(\epsilon)$$
$$\|f(t,\theta^1,\epsilon) - f(t,\theta^2,\epsilon)\| \leqq \Delta(\epsilon)\|\theta^1 - \theta^2\|$$
$$\|g(t,\theta^1,\epsilon) - g(t,\theta^2,\epsilon)\| \leqq \Delta(\epsilon)\|\theta^1 - \theta^2\|$$

for all $t \in E$, θ, θ^1, $\theta^2 \in E^k$, $0 < \epsilon \leqq \epsilon_1$, such that

$$x = f(t,\theta,\epsilon) \qquad y = g(t,\theta,\epsilon) \tag{15-13}$$

is an integral manifold of system (15-6). The behavior of the solutions on this integral manifold are obtained by solving the system

$$\begin{aligned} \dot{\theta} &= d(\epsilon) + \Theta(t,\theta,f(t,\theta,\epsilon),g(t,\theta,\epsilon),\epsilon) \\ \theta(t_0) &= \theta_0 \qquad \text{arbitrary} \end{aligned} \tag{15-14}$$

Only an outline of the proof of this theorem will be given. Conditions (15-10), (15-11) are slightly weaker than in Bogoliubov and Mitropolski [1], Hale [1], but the details of the proof are easily supplied by a slight modification of the procedure in Hale [1]. For the case where the functions Θ, X, Y in (15-6) are periodic in t, Theorem 15-1 has been proved by Diliberto [1].

Main Ideas of the Proof of Theorem 15-1 *and a Method of Approximating the Integral Manifold.* From (15-12), there is no loss in generality in assuming that $C = \operatorname{diag}(C_+,C_-)$, $A = \operatorname{diag}(A_+,A_-)$ where the eigenvalues of C_+, A_+ have positive real parts and the eigenvalues of C_-, A_- have negative real parts. Suppose D, Δ are fixed positive numbers, $D \leqq \min(\rho_1,\rho_2)$ and define the class of p-vector functions

$$\mathcal{C}_p(\Delta,D) = \{F(t,\theta,\epsilon) | F \in E^p,\ F(t,\theta,\epsilon) \text{ is multiply periodic in } \theta \text{ with vector period } \omega, \text{ is bounded by } D, \text{ and Lipschitzian in } \theta \text{ with Lipschitz constant } \Delta \text{ for all } t \in E,\ \theta \in E^k,\ 0 < \epsilon \leqq \epsilon_0\}$$

For any $F \in \mathcal{C}_m(\Delta,D)$, $G \in \mathcal{C}_n(\Delta,D)$, consider the equation

$$\dot{\theta} = d(\epsilon) + \Theta(t,\theta,F(t,\theta,\epsilon),G(t,\theta,\epsilon),\epsilon) \tag{15-15}$$

and let $\theta_t = T^{F,G}_{t-t_0,\,t_0}(\theta_0)$ be the solution (15-15) with initial value θ_0 at $t = t_0$.

With $J_C(t)$, $J_A(t)$ defined as in (13-4), for any $F \in \mathcal{C}_m(\Delta,D)$, $G \in \mathcal{C}_n(\Delta,D)$, define the transformation

$$\begin{aligned} S_{\theta,t}(F,G) &= [S^1_{\theta,t}(F,G),\ S^2_{\theta,t}(F,G)] \\ S^1_{\theta,t}(F,G) &= \epsilon \int_{-\infty}^{\infty} J_C(\epsilon u) X[u + t,\ T^{F,G}_{u,t}(\theta), \\ &\qquad F(u + t,\ T^{F,G}_{u,t}(\theta),\ \epsilon),\ G(u + t,\ T^{F,G}_{u,t}(\theta),\ \epsilon),\ \epsilon]\, du \\ S^2_{\theta,t}(F,G) &= \int_{-\infty}^{\infty} J_A(u) Y[u + t,\ T^{F,G}_{u,t}(\theta), \\ &\qquad F(u + t,\ T^{F,G}_{u,t}(\theta),\ \epsilon),\ G(u + t,\ T^{F,G}_{u,t}(\theta),\ \epsilon),\ \epsilon]\, du \end{aligned} \tag{15-16}$$

By first estimating the difference $T^{F^*,G^*}_{u,t_0}(\theta^*) - T^{F,G}_{u,t_0}(\theta^0)$ from (15-15), it is then shown that there exist an $\epsilon_1 > 0$ and functions $\Delta(\epsilon)$, $D(\epsilon)$, continuous in ϵ for $0 < \epsilon \leqq \epsilon_1$, $\Delta(\epsilon) \to 0$, $D(\epsilon) \to 0$ as $\epsilon \to 0$ such that the mapping $S_{\theta,t}(F,G)$ defined by (15-16) maps the set $\mathcal{C}_m(\Delta(\epsilon),\ D(\epsilon)) \times \mathcal{C}_n(\Delta(\epsilon),\ D(\epsilon))$ into itself and is a contraction. Consequently, there is a unique fixed point $f(t,\theta,\epsilon)$, $g(t,\theta,\epsilon)$ of this map. One then shows that this fixed point is an integral manifold of (15-6).

Since the map $S_{\theta,t}(F,G)$ is a contraction in the above class, the following iteration procedure will converge to the integral manifold for ϵ sufficiently small:

$$\begin{aligned}(F^0,G^0) &= 0 \\ (F^{k+1},G^{k+1}) &= S_{\theta,t}(F^k,G^k) \qquad k = 0, 1, 2, \ldots\end{aligned} \tag{15-17}$$

Lemma 15-1. If the conditions of Theorem 15-1 are satisfied and if there exists a sequence of real numbers $\{\tau_r\}$ such that, for each fixed ϵ, $0 < \epsilon \leqq \epsilon_0$, the functions Θ, X, Y in (15-6) satisfy the property

$$\begin{aligned}\|\Theta(t + \tau_r, \theta, x, y, \epsilon) - \Theta(t,\theta,x,y,\epsilon)\| &\to 0 \\ \|X(t + \tau_r, \theta, x, y, \epsilon) - X(t,\theta,x,y,\epsilon)\| &\to 0 \\ \|Y(t + \tau_r, \theta, x, y, \epsilon) - Y(t,\theta,x,y,\epsilon)\| &\to 0\end{aligned}$$

as $r \to \infty$ uniformly in (t,θ,x,y) for $t \in E$, $\theta \in E^k$, $0 \leqq \|x\| \leqq \rho_1$, $0 \leqq \|y\| \leqq \rho_2$, then for each fixed ϵ, $0 < \epsilon \leqq \epsilon_1$, the functions $f(t,\theta,\epsilon)$, $g(t,\theta,\epsilon)$ of Theorem 15-1 are such that

$$\|f(t + \tau_r, \theta, \epsilon) - f(t,\theta,\epsilon)\| \to 0 \qquad \|g(t + \tau_r, \theta, \epsilon) - g(t,\theta,\epsilon)\| \to 0$$

as $r \to \infty$ uniformly in t, θ for $t \in E$, $\theta \in E^k$.

The proof of this result is not too difficult if one uses the fact that f, g are fixed points of the operator S in (15-16).

Theorem 15-2. If the conditions of Theorem 15-1 are satisfied and for each fixed ϵ, $0 < \epsilon \leqq \epsilon_0$, the functions Θ, X, Y are almost periodic in t uniformly in θ, x, y, $\theta \in E^k$, $0 \leqq \|x\| \leqq \rho_1$, $0 \leqq \|y\| \leqq \rho_2$ (see Definition 12-1), then for each fixed ϵ, $0 < \epsilon \leqq \epsilon_1$, the functions f, g of Theorem 15-1 are almost periodic in t uniformly in θ, $\theta \in E^k$ with the same basic frequencies as Θ, X, Y.

The proof follows easily from Lemma 15-1 and the remarks concerning basic frequencies on page 114.

In particular, Theorem 15-2 implies that the functions $f(t,\theta,\epsilon)$, $g(t,\theta,\epsilon)$ are independent of t if the functions Θ, X, Y are independent of t. The functions $f(t,\theta,\epsilon)$, $g(t,\theta,\epsilon)$ are periodic in t of period T if the functions Θ, X, Y are periodic in t of period T.

For simplicity in stating the following theorem, the symbol $S_n(\rho)$ will designate the set of all n vectors z with $0 \leqq \|z\| \leqq \rho$.

Theorem 15-3. If the conditions of Theorem 15-1 are satisfied and s_1, s_2 of the eigenvalues of the matrices C, A, respectively, have negative real parts, then it is possible to find positive constants ϵ_1, γ, c, σ_0, σ_1 such that, for every ϵ in the half-open interval $(0,\epsilon_1]$ and each real $t_0 \in E$ and $\theta_0 \in E^k$, there exists, in the region $S_m(\sigma_0)$, an s_1-dimensional manifold $M_1(t_0,\theta_0,\epsilon)$ and, in the region $S_n(\sigma_0)$, an

s_2-dimensional manifold $M_2(t_0,\theta_0,\epsilon)$ which have the following properties, where $U_t = M_1(t,\theta_t,\epsilon) \times M_2(t,\theta_t,\epsilon)$:

1. If θ_t, x_t, y_t are the solutions of (15-6) with $\theta_{t_0} = \theta_0$, $x_{t_0} = x_0$, $y_{t_0} = y_0$, and if $(x_0,y_0) \in S_m(\sigma_0) \times S_n(\sigma_0) - U_{t_0}$, then for some $t^* > t_0$, $(x_{t^*},y_{t^*}) \notin S_m(\sigma_1) \times S_n(\sigma_1)$.
2. If $(x_0,y_0) \in U_{t_0}$, then for every $t \geqq t_0$,

$$\|x_t - f(t,\theta_t,\epsilon)\| \leqq c\{\exp[-\epsilon\gamma(t-t_0)]\}[\|x_0 - f(t_0,\theta_0,\epsilon)\| + \|y_0 - g(t_0,\theta_0,\epsilon)\|]$$
$$\|y_t - g(t,\theta_t,\epsilon)\| \leqq c\{\exp[-\gamma(t-t_0)]\}[\|x_0 - f(t_0,\theta_0,\epsilon)\| + \|y_0 - g(t_0,\theta_0,\epsilon)\|]$$

where f, g are the functions given in Theorem 15-1.

3. If the characteristic roots of the matrices C, A have positive real parts, then the manifolds $M_1(t_0,\theta_0,\epsilon)$, $M_2(t_0,\theta_0,\epsilon)$ consist of the points $x = f(t_0,\theta_0,\epsilon)$, $y = g(t_0,\theta_0,\epsilon)$, respectively.
4. If the real parts of the characteristic roots of C, A are negative, then the manifolds $M_1(t_0,\theta_0,\epsilon)$, $M_2(t_0,\theta_0,\epsilon)$ consist of the sets $S_m(\sigma_0)$, $S_n(\sigma_0)$, respectively.

For a proof of this theorem, see Bogoliubov and Mitropolski [1] or Hale [1]. The proof is similar to the proof of Theorem 13-1.

For systems of the form (15-6), Theorem 15-1 stated above gives some rather general conditions for the existence of integral manifolds, Theorem 15-2 discusses the behavior of this manifold as a function of the real variable t, and Theorem 15-3 asserts the stability properties in terms of the eigenvalues of the matrices C, A. We now wish to define a very general method of averaging for a particular class of systems and show that by a convenient transformation of variables one can reduce many problems to the discussion of systems of the form (15-6). The concept of averaging below has been used by many authors; for example, Krylov and Bogoliubov [1], Bogoliubov and Mitropolski [1], Diliberto [1] and Hale [1].

Suppose system (15-6) satisfies conditions (15-8) to (15-12) and in addition the function $X(t,\theta,x,y,\epsilon)$ has continuous first derivatives with respect to $(t,\theta,x,y,\epsilon) \in \Sigma_{\rho_1,\rho_2} \times (0,\epsilon_0]$ where Σ_{ρ_1,ρ_2} is defined in (15-7). We wish to consider the perturbed equations

$$\begin{aligned} \dot\theta &= d(\epsilon) + \Theta(t,\theta,x,y,\epsilon) + \epsilon\Theta^*(t,\theta,x,y,\epsilon) \\ \dot x &= \epsilon Cx + \epsilon X(t,\theta,x,y,\epsilon) + \epsilon X^*(t,\theta,x,y,\epsilon) \\ \dot y &= Ay + Y(t,\theta,x,y,\epsilon) \end{aligned} \tag{15-18}$$

where Θ^*, X^* are multiply periodic in θ of vector period ω, are continuous, and have continuous first partial derivatives with respect to t, θ, x, y for all $(t,\theta,x,y,\epsilon) \in \Sigma_{\rho_1,\rho_2} \times (0,\epsilon_0]$.

Theorem 15-4. If system (15-18) satisfies the above conditions, if the vectors θ, $d(\epsilon)$ are partitioned as $\theta = \text{col}\,(\varphi,s)$, $d = \text{col}\,(d_1,d_2)$ where φ, d_1 are l vectors, $d_1(0) = \text{col}\,(1, \ldots, 1)$, and if Θ^*, X^* have continuous second derivatives with respect to θ,x,

$$\begin{aligned} \Theta_0^*(t,\theta,x,y,\epsilon) &\overset{\text{def}}{\equiv} \lim_{T\to\infty} T^{-1} \int_0^T \Theta^*(t+\tau, \varphi+\tau, s, x, y, \epsilon)\, d\tau = 0 \\ X_0^*(t,\theta,x,y,\epsilon) &\overset{\text{def}}{\equiv} \lim_{T\to\infty} T^{-1} \int_0^T X^*(t+\tau, \varphi+\tau, s, x, y, \epsilon)\, d\tau = 0 \\ &\qquad \varphi + \tau = \text{col}\,(\varphi_1+\tau, \ldots, \varphi_l+\tau) \end{aligned} \tag{15-19}$$

then there exist an $\epsilon_1 > 0$ and an integral manifold of system (15-18) described parametrically by the equations

$$x = f(t,\theta,\epsilon) \qquad y = g(t,\theta,\epsilon) \qquad -\infty < t < \infty \qquad \theta \in E^k$$

where $0 < \epsilon \leqq \epsilon_1$, f, g satisfy the conditions of Theorems 15-1 to 15-3. If the mean value in (15-19) is taken only with respect to t (that is, the vector φ is absent), then X, Θ^*, X^* need not have derivatives with respect to t, θ.

Proof. Since the mean values Θ_0^*, X_0^* in (15-19) are zero, we can apply Lemmas 12-2 and Remark 12-1. These results imply, after a few simple calculations, that there exist functions $u(t,\theta,x,y,\epsilon)$, $v(t,\theta,x,y,\epsilon)$, multiply periodic in θ of vector period ω and, for every fixed ϵ, $0 < \epsilon \leqq \epsilon_1$ almost periodic in t uniformly in θ, x, y such that the transformation

$$\theta \to \theta + \epsilon u(t,\theta,x,y,\epsilon) \qquad x \to x + \epsilon v(t,\theta,x,y,\epsilon) \qquad y \to y$$

applied to (15-18) yields an equivalent set of differential equations of the form (15-6) which satisfies all the conditions of Theorems 15-1 to 15-3. The result then follows immediately.

The importance of Theorem 15-4 lies in the following fact: Let us call the "averaged" equations, the equations (15-18) with

$$\Theta^* = X^* = 0$$

Then the above theorem states that, if the averaged equations have an integral manifold of the type described above, so do the complete equations (15-18).

As mentioned before, Diliberto [1] has used a method of averaging as above for systems which are periodic in t. He has proved for this case some results more general than that above.

In the chapters which follow, we shall apply the results of this chapter to a large variety of problems.

16: Integral Manifolds—Noncritical Case

Let S be an $s + 1 \leqq (n + 1)$-dimensional integral manifold of the n-dimensional system

$$\dot{x} = X(x) \tag{16-1}$$

The manifold S is a set in $(n + 1)$-dimensional space. We say S is an *isolated integral manifold* if there exists a δ neighborhood U of S in $(n + 1)$-dimensional space such that, if S' is any other integral manifold of (16-1) and S' is in U, then S' is in S. If S is an isolated integral manifold, then one might suspect that, for ϵ sufficiently small, there is an isolated integral manifold S_ϵ, $0 \leqq \epsilon \leqq \epsilon_1$ (by isolated, we mean the δ neighborhood is independent of ϵ), of the equation

$$\dot{x} = X(x) + X^*(t,x,\epsilon) \tag{16-2}$$

if $X^*(t,x,\epsilon)$ is a uniformly bounded function of t for $-\infty < t < \infty$, $x \in U$, $0 \leqq \epsilon \leqq \epsilon_0$, $X^*(t,x,0) = 0$. Also, S_ϵ should approach S as $\epsilon \rightarrow 0$ and have the same stability properties as S. Unfortunately this is not the case in general, as the following simple example shows: Let x be a scalar and $X(x) = -x^3$, $X^*(t,x,\epsilon) = \epsilon x^2$. The line $x = 0$ is an isolated integral manifold of $\dot{x} = -x^3$ and is asymptotically stable, whereas the lines $x = 0$, $x = \epsilon$ are integral manifolds of $\dot{x} = -x^3 + \epsilon x^2$ and $x = 0$ is unstable while $x = \epsilon$ is stable.

The purpose of the present chapter is to discuss the case where S is an isolated integral manifold of (16-1) which has the property that there exists an isolated integral manifold S_ϵ, of (16-2) which has the same stability properties as S and $S_\epsilon \rightarrow S$ as $\epsilon \rightarrow 0$. This case is called *noncritical.*

We consider system (16-1) where x, X are n vectors and $X(x)$ is continuous together with its first and second partial derivatives with respect to x for $x \in U \subset E^n$. Suppose that system (16-1) has a nonconstant periodic solution $x^0(t)$ of period T, $x^0(t) \in U$ for all t. Furthermore, we suppose that $n - 1$ of the characteristic exponents of the linear variational equation

$$\dot{y} = \frac{\partial X\,(x^0(t))}{\partial x}\,y \tag{16-3}$$

have nonzero real parts. There is always one characteristic exponent of (16-3) which is equal to zero, since the periodic function $\dot{x}^0(t)$ is a solution of (16-3).

As we indicated in the first part of Chap. 15, the one-parameter family of periodic solutions of (16-1) given by $x^0(t + \varphi)$, φ an arbitrary constant, define a cylinder C in (x,t) space given parametrically by

$$C: x = x^0(\theta) \qquad 0 \leqq \theta \leqq T \qquad -\infty < t < \infty \tag{16-4}$$

Furthermore, the cylinder C is an integral manifold of system (16-1).

Theorem 16-1. If system (16-1) satisfies the conditions enumerated above and $X^*(t,x,\epsilon)$ is continuous and uniformly bounded together with its first and second partial derivatives with respect to x for $-\infty < t < \infty$, $x \in U$, $0 \leqq \epsilon \leqq \epsilon_0$, $X^*(t,x,0) = 0$, then there exist $\epsilon_0 > 0$, $\sigma_0 > 0$, such that for every ϵ, $0 < \epsilon \leqq \epsilon_0$, the following properties hold:

1. Equation (16-2) has an isolated integral manifold S_ϵ in the σ_0 neighborhood of the cylinder C in (16-4) and S_ϵ has a parametric representation

$$\begin{aligned} S_\epsilon: x &= f(t,\theta,\epsilon) \qquad 0 \leqq \theta \leqq T \qquad -\infty < t < \infty \\ f(t,\theta,0) &= x^0(\theta) \end{aligned} \tag{16-5}$$

where $f(t,\theta,\epsilon)$ is periodic in θ of period T and is bounded and uniformly continuous together with its derivatives with respect to θ up through order 2. Furthermore, there exists a scalar function $F(t,\theta,\epsilon)$ periodic in θ of period T, possessing bounded and uniformly continuous derivatives with respect to θ up through order 2 such that system (16-2) is equivalent to the equation

$$\frac{d\theta}{dt} = 1 + F(t,\theta,\epsilon) \tag{16-6}$$

on the manifold S_ϵ.

2. If for each fixed ϵ, $0 \leqq \epsilon \leqq \epsilon_0$, $X^*(t,x,\epsilon)$ is almost periodic in t uniformly with respect to $x \in U$, then, for each fixed ϵ, the functions

$f(t,\theta,\epsilon)$ are almost periodic in t uniformly with respect to θ and have the same basic frequencies as $X^*(t,x,\epsilon)$.

3. If the integral manifold C of system (16-1) is stable, unstable, or conditionally stable with respect to a manifold of dimension s,† then the integral manifold S of system (16-2) is stable, unstable, or conditionally stable with respect to a manifold of dimension s.

The proof of this theorem proceeds as follows: It is known (see, for example, Diliberto and Hufford [1], Urabe [1]) that there exists a transformation taking x into (θ,y), θ a scalar, and y an $(n - 1)$ vector, which is periodic in θ of period T, and one-to-one in a neighborhood of the closed curve $x = x^0(\theta)$, $0 \leqq \theta \leqq T$, such that the system (16-2) is equivalent to the system

$$\dot{\theta} = 1 + \Theta_1(\theta,y) + \Theta_2(t,\theta,y,\epsilon)$$
$$\dot{y} = Ay + Y_1(\theta,y) + Y_2(t,\theta,y,\epsilon)$$

$\Theta_1 = O(||y||)$, $Y_1 = O(||y||^2)$ as $||y|| \to 0$, $\Theta_2 = 0$, $Y_2 = 0$ for $\epsilon = 0$, and the eigenvalues of the constant matrix A have nonzero real parts. This system is now a special case of system (16-6) with the vector x absent and satisfies conditions (16-8) to (16-12). The theorem then follows from the results of Chap. 15.

In case $X^*(t,x,\epsilon)$ in (16-2) is periodic in t, then the function $F(t,\theta,\epsilon)$ in (16-6) is also periodic in t with the same period. Consequently, it follows from the theory of systems of the form (16-6) (see Coddington and Levinson [1]) that the solutions of (16-4) on the integral manifold S_ϵ are either almost periodic functions, periodic functions, or approach periodic functions as $t \to \infty$. Theorem 16-1 for this special case was proved by Levinson [1].

Theorem 16-2. Suppose $X(x)$ satisfies the conditions of Theorem 16-1 and let $X^*(\tau,x)$ be continuous and uniformly bounded together

† Suppose $\dot{x} = X(t,x)$ is an nth-order differential system and suppose that S is an integral manifold of this equation of dimension m and has a parametric representation $x = f(t,\theta)$, $\theta \in E^m$. U with a subscript will always designate a neighborhood of S. The manifold S will be said to be *stable* if for every U_2 there is a U_1 such that, for every solution $x(t)$ of the above equation with $[x(t_0),t_0] \in U_1$, we have $[x(t),t] \in U_2$ for all $t \geqq t_0$ and dist $[(x(t),t),S] \to 0$ as $t \to \infty$. If there is a U_2 such that, for every $U_1 \subset U_2$ and every solution $x(t)$ with $[x(t_0),t_0] \in U_1 - S$, there exists a $t^* > t_0$ such that $[x(t^*),t^*] \notin U_2$, then S will be said to be *unstable*. If there is a set M_1 of dimension $s < n + 1$ such that for every U_2 there is a U_1 such that for every solution $x(t)$ with $[x(t_0),t_0] \in U_1 \cap M_1$, we have $[x(t),t] \in U_2$ for all $t \geqq t_0$ and dist $[(x(t),t),S] \to 0$ as $t \to \infty$, and there is a U_3 such that, for every $U_1 \subset U_3$ and $[x(t_0),t_0] \in U_1 - M_1$, there must be a $t^* > t_0$ such that $[x(t^*),t^*] \notin U_3$, then S will be said to be *conditionally stable with respect to a manifold of dimension* s.

with its first and second partial derivatives with respect to x for $-\infty < \tau < \infty$, $x \in U$. If $X^*(\tau,x)$ is almost periodic in τ uniformly in x, $x \in U$ and

$$\lim_{T\to\infty} T^{-1} \int_0^T X^*(\tau,x)\, d\tau = 0 \tag{16-7}$$

then there exist $\omega_0 > 0$, $\sigma_0 > 0$ such that, for every $\omega \geqq \omega_0$, the equation

$$\dot{x} = X(x) + X^*(\omega t,x) \tag{16-8}$$

has a unique integral manifold S_ω in the σ_0 neighborhood of the cylinder C in (16-4), $S_\omega \to C$ as $\omega \to \infty$, and S_ω has a parametric representation

$$S_\omega\colon x = f(\omega t,\theta,\omega^{-1}) \qquad 0 \leqq \theta \leqq T \qquad -\infty < t < \infty$$

where f is periodic in θ of period T and, for each fixed $\omega \geqq \omega_0$, is almost periodic in T uniformly in θ, $\theta \in R$, with the same basic frequencies as $X^*(\omega t,x)$. Finally, the stability properties of S_ω are the same as those of C.

Proof. From (16-7) and Lemma 12-2, for any $\eta > 0$, there exists a function $w(\tau,x,\eta)$ such that

$$\left\| \frac{\partial w\,(\tau,x,\eta)}{\partial\tau} - X^*(\tau,x) \right\| \leqq \sigma(\eta) \tag{16-9}$$

for all τ, x and $\sigma(\eta) \to 0$ as $\eta \to 0$. Furthermore, from Lemma 12-2, if $\eta = \omega^{-1}$, $\omega \geqq \omega_0$, ω_0 sufficiently large, the transformation

$$x = y + \frac{1}{\omega} w\left(\omega t, y, \frac{1}{\omega}\right)$$

is one-to-one and, if applied to (16-8), yields

$$\left[I + \frac{1}{\omega}\frac{\partial w\left(\omega t,y,\frac{1}{\omega}\right)}{\partial y}\right]\dot{y} = X(y) + \left[X\left(y + \frac{1}{\omega}w\right) - X(y)\right]$$
$$+ \left[X^*(\omega t,y) - \frac{\partial w\left(\omega t,y,\frac{1}{\omega}\right)}{\partial\tau}\right] + \left[X^*\left(\omega t, y + \frac{1}{\omega}w\right) - X^*(\omega t,y)\right]$$

But, from (16-9), and the fact that

$$\left[I + \frac{1}{\omega}\frac{\partial w}{\partial y}\right]^{-1} = I - \frac{1}{\omega}\frac{\partial w}{\partial y} + \cdots$$

for ω sufficiently large, it follows that

$$\dot{y} = X(y) + G\left(\omega t, y, \frac{1}{\omega}\right)$$

where $G(\tau,y,\omega^{-1})$ is continuous in τ, y, ω and uniformly bounded for $-\infty < \tau < \infty$, $x \in U$, $\omega \geqq \omega_0$, $G(\tau,y,0) = 0$.

As in the proof of Theorem 16-1, there exists a transformation taking y into (θ,z) where θ is a scalar and z an $(n - 1)$ vector, which is periodic in θ of period T, and one-to-one in a neighborhood of the closed curve $y = x^0(\theta)$, $0 \leqq \theta \leqq T$, such that our system is equivalent to

$$\begin{aligned} \dot{\theta} &= 1 + \Theta_1(\theta,z) + \Theta_2(\omega t,\theta,z,\omega^{-1}) \\ \dot{z} &= Az + Z_1(\theta,z) + Z_2(\omega t,\theta,z,\omega^{-1}) \end{aligned}$$

$\Theta_1 = O(\|z\|)$, $Z_1 = O(\|z\|^2)$ as Z_1 $(\|z\|^2)$ as $\|z\| \to 0$, $\Theta_2(\tau,\theta,z,0) = 0$, $Z_2(\tau,\theta,z,0) = 0$, and the eigenvalues of the matrix A have nonzero real parts. If $\omega t = \tau$, this system becomes

$$\begin{aligned} \frac{d\theta}{d\tau} &= \omega^{-1}[1 + \Theta_1(\theta,z) + \Theta_2(\tau,\theta,z,\omega^{-1})] \\ \frac{dz}{d\tau} &= \omega^{-1}[Az + Z_1(\theta,z) + Z_2(\tau,\theta,z,\omega^{-1})] \end{aligned}$$

Since this system is a special case of (15-6) with $\omega^{-1} = \epsilon$, $x = z$ and the variable y absent, Theorem 16-2 follows from Theorems 15-1, 15-2, and 15-3.

It should be clear that a more general result which includes both Theorems 15-1 and 15-2 could be stated; that is, one could have two parameters ϵ and ω, ϵ being chosen small and ω large. We may use this more general result in examples.

Example 16-1. *Forced van der Pol Equation.* An example illustrating the above theorems is the forced van der Pol equation

$$\begin{aligned} \dot{x}_1 &= x_2 \\ \dot{x}_2 &= -x + k(1 - x_1^2)x_2 + \epsilon A \sin \omega_1 t + \epsilon B \sin \omega_2 t \end{aligned} \tag{16-10}$$

where $k > 0$, $\epsilon > 0$, $\omega_1 > 0$, $\omega_2 > 0$, A, B are constants and x_1, x_2 are scalars. It is known that the autonomous equation

$$\dot{x}_1 = x_2 \qquad \dot{x}_2 = -x_1 + k(1 - x_1^2)x_2 \tag{16-11}$$

has an asymptotically stable limit cycle of period $T(k)$ for every $k > 0$ and the corresponding linear variational equation (16-3) has one characteristic exponent which is negative. If we let C designate the cylinder generated by the periodic solutions of (16-11), then Theorem 16-1 asserts the existence of a generalized cylinder C_ϵ for ϵ sufficiently small, defined parametrically by

$$C_\epsilon: x_1 = f_1(t,\theta,\epsilon) \quad x_2 = f_2(t,\theta,\epsilon) \quad 0 \leqq \theta \leqq T \quad -\infty < t < \infty \qquad (16\text{-}12)$$

where f_1, f_2 are almost periodic in t with basic frequencies ω_1, ω_2 and periodic in θ of period $T(k)$. Furthermore, $C_\epsilon \to C$ as $\epsilon \to 0$ and C_ϵ is asymptotically stable. If $\omega_2 = c\omega_1$ where c is a constant, then Theorem 16-2 asserts the existence of a generalized cylinder $C_{\omega_1^{-1}}$, ω_1 sufficiently large, having a parametric representation of the form (16-12) with ϵ replaced by ω_1^{-1} and $C_{\omega_1^{-1}} \to C$ as $\omega_1 \to \infty$ and $C_{\omega_1^{-1}}$ is asymptotically stable. For this particular example, with $B = 0$, the cross sections of the cylinder C_ϵ have been calculated numerically by Loud [4].

Example 16-2. *Coupled van der Pol Equations.* As another example, consider two coupled van der Pol equations

$$\begin{aligned} \dot{x}_1 &= x_2 \\ \dot{x}_2 &= -x_1 + k_1(1 - x_1^2)x_2 + \epsilon f(t, x_1, \ldots, x_4) \\ \dot{x}_3 &= x_4 \\ \dot{x}_4 &= -\sigma^2 x_3 + k_2(1 - x_3^2)x_4 + \epsilon g(t, x_1, \ldots, x_4) \end{aligned} \qquad (16\text{-}13)$$

where x_1 to x_4 are scalars and k_1, k_2, σ are constants, and f, g are given functions sufficiently smooth in t, x, y and almost periodic in t uniformly with respect to x, y in some set.

For $\epsilon = 0$, these equations are uncoupled in the sense that the first two equations are independent of the last two. Also, if $x = \text{col}\,(x_1,x_2)$, $y = \text{col}\,(x_3,x_4)$, then we may write these equations, (16-13), as

$$\begin{aligned} \dot{x} &= X(x) + \epsilon X^*(t,x,y) \\ \dot{y} &= Y(y) + \epsilon Y^*(t,x,y) \end{aligned} \qquad (16\text{-}14)$$

where X, Y, X^*, Y^* are 2 vectors, which are easily written down from (16-13). From the nature of the van der Pol equations,

$$\dot{x} = X(x) \qquad (16\text{-}15)$$

$$\dot{y} = Y(y) \qquad (16\text{-}16)$$

if σ, k_1, k_2 are >0, then there exist a periodic solution $x = x^0(t)$ of (16-15) of period T_1, and a periodic solution $y = y^0(t)$ of (16-16) of period T_2, which are exponential asymptotically orbitally stable and the corresponding linear variational equations have a characteristic exponent with negative real part. Also, if 0_x, 0_y designate the origins in x, y space, respectively, then the coefficient matrix of the linear variational equations for the solutions $x = 0_x$, $y = 0_y$ of (16-15), (16-16) have eigenvalues with positive real parts. Consequently, if we let C_{x^0}, C_{y^0} designate the cylinders defined by $x = x^0(\theta)$, $0 \leqq \theta \leqq T_1$, $y = y_0(\theta)$, $0 \leqq \theta \leqq T_2$, $-\infty < t < \infty$, respectively, then there are four types of integral manifolds in (x,y,t) space of system (16-14) for $\epsilon = 0$ which are of interest:

Case 1. The unstable line $0_x \times 0_y$
Case 2. The unstable manifold $S_2 = C_{x^0} \times 0_y$
Case 3. The unstable manifold $S_3 = 0_x \times C_{y^0}$
Case 4. The stable manifold $S_4 = C_{x^0} \times C_{y^0}$

We wish now to discuss the types of behavior which one expects for the perturbed equation (16-14) for ϵ sufficiently small.

Case 1. If $f(t,0,0)$, $g(t,0,0)$ are not both identically zero then the unstable line generates an unstable almost periodic motion of (16-14) for ϵ sufficiently small and this motion degenerates to $0_x \times 0_y$ as $\epsilon \to 0$. This is fairly obvious since we can apply Theorems 15-1 to 15-3 to our system (16-13) which is a special case of (15-6) with the variables θ, x absent and y replaced by $y = \text{col}\,(x_1, \ldots, x_4)$. If $f(t,0,0) = 0$, $g(t,0,0) = 0$, then $0_x \times 0_y$ is a solution of (16-14).

Case 2. One can introduce new polar-type coordinates as in the proof of Theorem 16-1 in a neighborhood of the cylinder C_{x^0} by replacing (x_1,x_2) by variables (θ,h) where θ, h are scalars to obtain in place of (16-13) a system of the form

$$\begin{aligned}
\dot{\theta} &= 1 + \Theta_1(\theta,h) + \epsilon\Theta_2(t,\theta,h,x_3,x_4) \\
\dot{h} &= ah + H_1(\theta,h) + \epsilon H_2(t,\theta,h,x_3,x_4) \\
\dot{x}_3 &= x_4 \\
\dot{x}_4 &= -\sigma^2 x_3 + k_2 x_4 - k_2 x_3^2 x_4 + \epsilon G(t,\theta,h,x_3,x_4)
\end{aligned}$$

where $a < 0$, $\Theta_1, = O(\|h\|)$, $H_1 = O(\|h\|^2)$ as $\|h\| \to 0$. Theorems 15-1 15-3 may again be applied to this system which is a special case of to (15-6) with the variable x absent and y replaced by (h,x_3,x_4). Thus there exists an unstable integral manifold of (16-13), $S_{2\epsilon}$, $S_{2\epsilon} \to S_2$ as $\epsilon \to 0$, and the parametric representation of this manifold is almost periodic in t.

Case 3. One treats this case in the same way as case 2.

Case 4. In this case, one introduces new polar-type coordinates $(x_1,x_2) \to (\theta_1,h)$, $(x_3,x_4) \to (\theta_2,h_2)$, in a neighborhood of the closed orbits of the periodic solutions of the unperturbed equations to obtain

$$\begin{aligned} \dot{\theta}_j &= 1 + \Theta_{j1}(\theta_j,h_j) + \epsilon\Theta_{j2}(t,\theta_1,\theta_2,h_1,h_2) \\ \dot{h}_j &= -a_j h_j + H_{j1}(\theta_j,h_j) + \epsilon H_{j2}(t,\theta_1,\theta_2,h_1,h_2) \qquad j = 1, 2 \end{aligned}$$

where $a_j > 0$, H_{11}, $H_{21} = O(\|h\|^2)$ as $\|h\| \to 0$, $j = 1, 2$. Theorems 15-1 to 15-3 may again be applied to this system which is a special case of (15-6) with the variable x absent and y replaced by (h_1,h_2). Thus there exists a stable integral manifold of (16-13), $S_{4\epsilon}$, $S_{4\epsilon} \to S_4$ as $\epsilon \to 0$ and the parametric representation of this manifold is almost periodic in t.

The discussion of the example above could clearly be generalized, and the interested reader may look in Kyner [1] and Hale [1].

Example 16-3. *Quenching of Oscillations.* Suppose that the n-dimensional autonomous system

$$\dot{x} = X(x) \tag{16-17}$$

where X has continuous second derivatives up through order 2 in x for $x \in U$, has a limit cycle $x^0(t)$ such that the corresponding variational equation has $(n - 1)$ characteristic exponents with negative real parts. It may be that the oscillation $x^0(t)$ is undesirable. We ask the following question: Given $\epsilon > 0$, is it possible to find an n-vector function $f_\epsilon(t)$ such that the system

$$\dot{x} = X(f_\epsilon(t) + x) \tag{16-18}$$

has no bounded integral manifold which deviates from the line $x = 0$ by more than ϵ? If the answer to this question is in the affirmative, we say the oscillation $x^0(t)$ has been *quenched*. Theorem 16-2 above can be useful in solving this problem. In fact, let $\omega > 0$ be given and let $g(\tau,B)$ be an n-vector periodic function of period T depending upon some parameters $B = (B_1, \ldots, B_N)$, $g(\tau,0) = 0$, and consider the equation

$$\begin{aligned} \dot{x} &= X_0(B,x) + [X\{g(\omega t,B) + x\} - X_0(B,x)] \\ X_0(B,x) &= \frac{1}{T}\int_0^T X\{g(\tau,B) + x\}\, d\tau \end{aligned} \tag{16-19}$$

It is clear that $X_0(B,x)$ is independent of ω. For some fixed B, suppose that the averaged equations

$$\dot{x} = X_0(B,x) \tag{16-20}$$

have a limit cycle $x_0(t,B)$ whose associated linear variational equation has $(n-1)$ characteristic exponents with nonzero real parts, and let C_B be the cylinder in x, t generated by these periodic solutions. From Theorem 16-2, there exists an $\omega_0 = \omega_0(B)$ and a generalized cylinder $C_{B,\omega}$, $\omega \geqq \omega_0$, such that the parametric representation of $C_{B,\omega}$ is periodic in t of period T/ω and $C_{B,\omega}$ is an integral manifold of (16-19), $C_{B,\omega} \to C_B$ as $\omega \to \infty$ and $C_{B,\omega}$ has the same stability properties as C_B.

Since $g(\omega t,0) = 0$, it follows that $X_0(0,x) = X(x)$ and thus

$$x_0(t,0) = x^0(t)$$

where $x^0(t)$ is the limit cycle of (16-17). Consequently, from the remarks made above, if, for example, $X(x)$ is of such a nature that the periodic solution $x_0(t,B)$ of (16-20) satisfies $\|x_0(t,B)\| \to 0$ as $\|B\| \to \infty$, then for every $\epsilon > 0$ there exists a $B_1(\epsilon)$ such that $\|x_0(t,B_0)\| \leqq \epsilon/2$ and, if $\omega \geqq \omega_0(\epsilon)$ sufficiently large, then the distance from the integral manifold $C_{B,\omega}$ to the line $x = 0$ is $<\epsilon$ and the oscillation $x^0(t)$ is quenched.

As an example of this type of quenching, let us consider the van der Pol equation

$$\begin{aligned} \dot{x}_1 &= x_2 \\ \dot{x}_2 &= -x_1 + k(1 - x_1^2)x_2 \qquad k > 0 \end{aligned} \tag{16-21}$$

and the related system

$$\begin{aligned} \dot{x}_1 &= x_2 \\ \dot{x}_2 &= -x_1 + k[1 - (x_1 + B \sin \omega t)^2]x_2 \end{aligned} \tag{16-22}$$

where B, ω are parameters. The averaged equations associated with (16-22) are

$$\begin{aligned} \dot{x}_1 &= x_2 \\ \dot{x}_2 &= -x_1 + k\left(1 - \frac{B^2}{2}\right)\left[1 - \frac{x_1^2}{1 - (B^2/2)}\right]x_2 \end{aligned} \tag{16-23}$$

It is known that (16-23) has a self-excited oscillation for $B^2 < 2$ [notice that $B = 0$ gives system (16-21)], and the corresponding linear variational equations have a negative characteristic exponent. Consequently, for every $B^2 < 2$ there exists an $\omega_0 = \omega_0(B)$ such that there exist a stable integral manifold of (16-22) for every $\omega \geqq \omega_0$ and this integral manifold approaches, as $\omega \to \infty$, the cylinder generated by the self-excited oscillation of (16-23). But, for B^2 very close to 2, the amplitude of this oscillation is easily determined by the methods outlined in Chap. 10. In fact, if $B^2 < 2$ and we let $x_1 = y_1\sqrt{1 - B^2/2}$,

$x_2 = y_2\sqrt{1 - B^2/2}$, then

$$\begin{aligned}\dot{y}_1 &= y_2 \\ \dot{y}_2 &= -y_1 + k\left(1 - \frac{B^2}{2}\right)(1 - y_1^2)y_2\end{aligned} \tag{16-24}$$

For B^2 close to 2, we can let $\epsilon = 1 - B^2/2$ and obtain from Chap. 10 or Sec. 7-5 that the self-excited oscillation of (16-23) has an amplitude approximately equal to $2\sqrt{1 - B^2/2}$ for B^2 close enough to 2 and the oscillation is quenched according to our definition above.

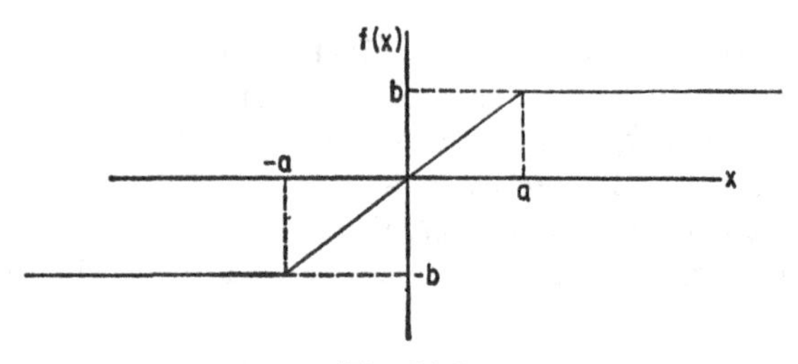

Fig. 16-1

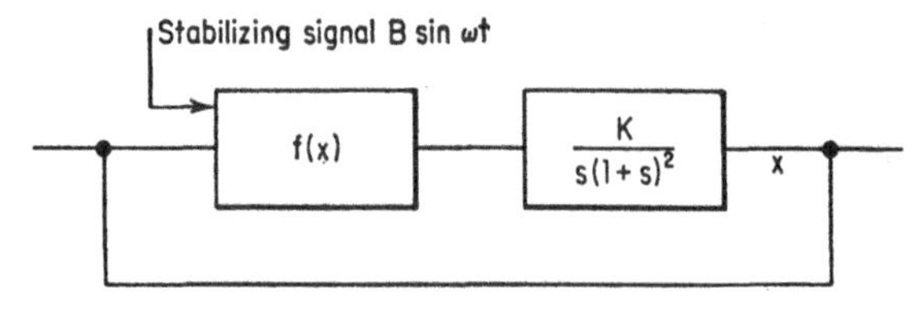

Fig. 16-2

Another type of quenching may occur by the introduction of the internal forcing $g(\omega t, B)$ in (16-19). We illustrate this by means of an example which was investigated by Boyer [1] (see also Oldenburger and Boyer [1]) and later give some remarks about how this type of quenching may be investigated mathematically.

Consider the third-order differential equation

$$\dddot{x} + 2\ddot{x} + \dot{x} + Kf(x) = 0 \tag{16-25}$$

where $f(x)$ is given in Fig. 16-1. System (16-25) for some values of K has a self-excited oscillation which is asymptotically stable, and the problem is to try to quench this oscillation by replacing $f(x)$ by $f(x + B\sin\omega t)$ and choosing B, ω large. The block diagram for such a system is given in Fig. 16-2.

The results of Boyer are given in Table 16-1 with $b = 10$, $a = 5$ in Fig. 16-1, and ω chosen to be at least twenty times the frequency of the self-excited oscillation of (16-25).

Table 16-1

K	B	Amplitude of oscillation
2	0	13.2
	5	12.2
	7	10.0
	7.5	8.6
	8	0.0
	10	0.0
4	0	26.4
	5	26.3
	10	25.5
	15	23.0
	15.5	22.0
	16	21.0
	17	0.0
	20	0.0

As we see from Table 16-1, for sufficiently large values of ω and B the oscillation disappears. More specifically, the oscillation disappears at a finite value of B, and furthermore, before its disappearance, the amplitude of the oscillation was very large.

To obtain an intuitive feeling for why such a phenomenon occurs, let us investigate this problem by using the method of averaging. We wish to determine the oscillatory character of an equation

$$\dddot{x} + 2\ddot{x} + \dot{x} + Kf(x + B \sin \omega t) = 0 \tag{16-26}$$

for large values of B, ω. Let

$$f_0(x,B) = \frac{1}{2\pi} \int_0^{2\pi} f(x + B \sin \tau)\, d\tau \tag{16-27}$$

and consider the averaged equation

$$\dddot{x} + 2\ddot{x} + \dot{x} + Kf_0(x,B) = 0 \tag{16-28}$$

By using the transformation indicated in the proof of Theorem 16-2, one observes that, for ω sufficiently large, the averaged equation (16-28) is indeed a good approximation to the differential system (16-26).

More specifically, if system (16-28) has a self-excited oscillation with two of the corresponding characteristic exponents with nonzero real parts, then for ω large there is an oscillation of (16-26) close to the one of (16-28). What is the nature of the nonlinear characteristic $f_0(x,B)$ in (16-28)? Without too much difficulty, one observes that they have essentially the same shape as $f(x)$ in Fig. 16-1, except the saturation point a is replaced by $a + B$ (see Fig. 16-3). Consequently, one would suspect that, if B is larger than some value B_0, then no self-excited oscillation of (16-28) will occur and the zero solution of (16-28)

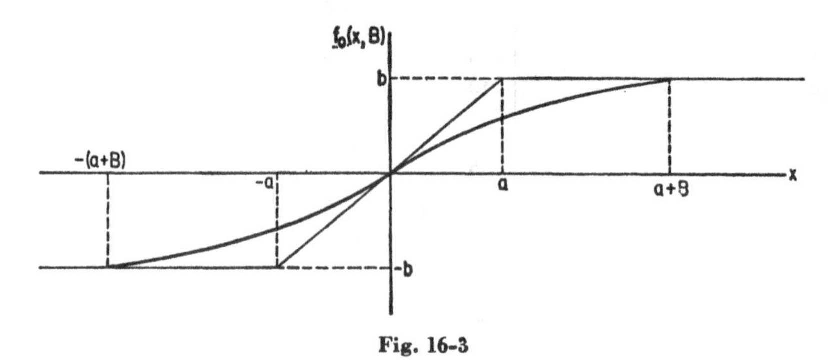

Fig. 16-3

is asymptotically stable with respect to all initial values (asymptotically stable in the large).

On the other hand, this does not begin to explain why the oscillation had a very large amplitude just before it disappeared. To explain this phenomenon, Boyer used the method of describing functions to analyze (16-28) [the describing function for $f_0(x,B)$ is called by Boyer a pseudo-describing function for (16-26)]. The concept of a describing function was explained in Sec. 11-2. As we have indicated, it is very difficult to justify this procedure mathematically, but practically it gives in most cases very useful information about the behavior of the solutions.

Applying describing functions, Boyer concluded that, before the disappearance of the stable self-excited oscillation of (16-28), the system also had an unstable oscillation which was very close to the stable one. After the oscillation has disappeared, the zero solution of (16-28) becomes asymptotically stable in the large. In this case, it does not necessarily follow that all solutions of (16-26) approach some neighborhood U_ω of the origin $x = \dot{x} = \ddot{x} = 0$, $t \to \infty$, where U_ω can be made arbitrarily small by taking ω large. On the other hand, if one assumes the asymptotic stability of the zero solution of (16-28) is strong enough, then by using the theory of Liapunov functions (a subject which would take us too far afield in this monograph), one can show that, for any given r, there exist an ω_0 and a δ_ω, $\omega \geqq \omega_0$, $\delta_\omega \to 0$ as $\omega \to \infty$, such that, for any $\omega \geqq \omega_0$, any solution of (16-26) with initial value $\|x_0\| \leqq r$, $\|\dot{x}_0\| \leqq r$, $\|\ddot{x}_0\| \leqq r$, will lie in the region $\|x\| \leqq \delta_\omega$, $\|\dot{x}\| \leqq \delta_\omega$, $\|\ddot{x}\| \leqq \delta_\omega$ after some finite time.

There seem to be many interesting and unanswered questions concerning the problem of quenching of oscillations. Even though we cannot prove rigorously too much about this problem, it seems as if the idea of embedding the differential system (16-17) into a class of differential systems (16-19) which contain the vector parameter B may prove to be useful and deserves further mathematical investigation.

17: Almost Periodic Solutions—Critical Case

In this chapter, we illustrate some easy consequences of Theorem 15-4. These results will be applied to the study of almost periodic solutions of nonlinear almost periodic differential equations in critical cases (see Chap. 13) and to studying the stability of the zero solution of a linear almost periodic system.

Consider the system of equations

$$\begin{aligned} \dot{x} &= \epsilon X(t,x,y,\epsilon) \\ \dot{y} &= Ay + \epsilon Y(t,x,y,\epsilon) \end{aligned} \tag{17-1}$$

where ϵ is a real parameter, x, X are m vectors, y, Y are n vectors, A is a constant $n \times n$ matrix, X, Y are bounded and uniformly continuous in t, x, y, ϵ and have bounded and uniformly continuous first and second partial derivatives with respect to x, y for $-\infty < t < \infty$, $0 \leqq \|x\| \leqq R_1$, $0 \leqq \|y\| \leqq R_2$, $0 \leqq \epsilon \leqq \epsilon_0$. Furthermore, for each fixed ϵ, $0 \leqq \epsilon \leqq \epsilon_0$, we suppose that X, Y are almost periodic in t uniformly with respect to x, y for $0 \leqq \|x\| \leqq R_1$, $0 \leqq \|y\| \leqq R_2$, and the matrix A has all eigenvalues with nonzero real parts.

Theorem 17-1. If system (17-1) satisfies the conditions enumerated above, if we let

$$X_0(x,y,\epsilon) = \lim_{T \to \infty} T^{-1} \int_0^T X(t,x,y,\epsilon)\, dt \tag{17-2}$$

and if there exists a constant m vector x^0 such that $X_0(x^0,0,0) = 0$ and the eigenvalues of the matrix $\partial X(x^0,0,0)/\partial x$ have nonzero real parts, then it is possible to find positive constants ϵ_1, σ_0 and vector functions $f(t,\epsilon)$, $g(t,\epsilon)$, of dimensions m, n, respectively, such that $f(t,\epsilon)$, $g(t,\epsilon)$ are continuous in t, ϵ, $-\infty < t < \infty$, $0 < \epsilon \leqq \epsilon_1$; for each fixed ϵ, $0 <$

$\epsilon \leq \epsilon_1$, almost periodic in t (with the same basic frequencies as X, Y), $f(t,0) = x^0$, $g(t,0) = 0$, and $x = f(t,\epsilon)$, $y = g(t,\epsilon)$ is a solution of (17-1). Furthermore, for $\epsilon \neq 0$, this is the only solution of (17-1) that remains in the region $0 \leqq \|x - x^0\| \leqq \sigma_0$, $0 \leqq \|y\| \leqq \sigma_0$ for all t, $-\infty < t < \infty$. The stability properties of the solution $x = f(t,\epsilon)$, $y = g(t,\epsilon)$ are locally the same as the stability properties of the solution $x = 0$, $y = 0$ of the system

$$\dot{x} = \frac{\partial X\,(x^0,0,0)}{\partial x}\,x \qquad \dot{y} = Ay$$

Proof. If we rewrite (17-1) as

$$\begin{aligned} \dot{x} &= \epsilon X_0(x,y,0) + \epsilon[X_0(x,y,\epsilon) - X_0(x,y,0)] + \epsilon[X(t,x,y,\epsilon) - X_0(x,y,\epsilon)] \\ \dot{y} &= Ay + \epsilon Y(t,x,y,\epsilon) \end{aligned}$$

and make the transformation

$$x \to x_0 + x + \epsilon\,\frac{\partial X\,(x^0,0,0)}{\partial y}\,A^{-1}y \qquad y \to y$$

then the new equation has the form

$$\begin{aligned} \dot{x} &= \epsilon\,\frac{\partial X\,(x^0,0,0)}{\partial x}\,x + \epsilon X_1(t,x,y,\epsilon) + \epsilon X^*(t,x,y,\epsilon) \\ \dot{y} &= Ay + \epsilon Y_1(t,x,y,\epsilon) \end{aligned}$$

where X^* has mean value zero with respect to t and X_1, ϵY_1 satisfy the same conditions as X, Y in (15-18). The result then follows from Theorem 15-4 with the vector θ absent.

Theorem 17-2. If system (17-1) satisfies all the conditions enumerated above, if $X_0(x,y,\epsilon)$ is defined by (17-2) and there exists a periodic solution $x^0(\epsilon\omega t)$, $x^0(s + 2\pi) = x^0(s)$, of the equation

$$\dot{x} = \epsilon X_0(x,0,0)$$

such that the linear variational equation has $m - 1$ characteristic exponents with nonzero real parts, then there exist positive numbers ϵ_1, σ_0, such that for every ϵ, $0 < \epsilon \leqq \epsilon_1$, the following conclusions hold:

1. Equation (17-1) has a unique integral manifold S lying for all real t in a σ_0 neighborhood of the cylinder $x = x^0(\theta)$, $y = 0$, $0 \leqq \theta \leqq 2\pi$, $-\infty < t < \infty$ and has a parametric representation of the form

$$x = f(t,\theta,\epsilon) \qquad y = g(t,\theta,\epsilon) \qquad 0 \leqq \theta \leqq 2\pi \qquad -\infty < t < \infty$$

where f, g are periodic in θ of period 2π, almost periodic in t uniformly with respect to $\theta \in R$ for each fixed ϵ, $0 < \epsilon \leqq \epsilon_1$, have uniformly continuous derivatives with respect to θ up through order 2, $\|f(t,\theta,\epsilon) - x^0(\theta)\| \leqq \delta(\epsilon)$, $\|g(t,\theta,\epsilon)\| \leqq \delta(\epsilon)$, $\delta(\epsilon) \to 0$ as $\epsilon \to 0$, $-\infty < t < \infty$, $0 \leqq \theta \leqq 2\pi$.

2. If the cylinder $x = x^0(\theta)$, $y = 0$, $0 \leqq \theta \leqq 2\pi$, $-\infty < t < \infty$, which is an integral manifold of

$$\dot{x} = \epsilon X(x,0,0) \qquad \dot{y} = Ay$$

is stable, unstable, or conditionally stable with respect to a manifold of dimension s, then the integral manifold S of (17-1) is stable, unstable, or conditionally stable with respect to a manifold of dimensions s.

The proof of this theorem follows easily from Theorem 15-4 if one uses the transformation mentioned in the proof of Theorem 16-1.

Let us now apply Theorem 17-1 to determine conditions which will ensure that the $(n + m)$-dimensional system

$$\dot{z} = Dz + \epsilon q(t,z,\epsilon) \tag{17-3}$$

has almost periodic solutions for ϵ sufficiently small. We suppose, for each fixed ϵ, $0 \leqq \epsilon \leqq \epsilon_0$, that q is almost periodic in t uniformly with respect to z for $0 \leqq \|z\| \leqq R$, $R > 0$, is continuous in t, z, ϵ, and has continuous first and second partial derivatives with respect to z for $-\infty < t < \infty$, $0 \leqq \epsilon \leqq \epsilon_0$. Also, the matrix D is such that

$$D = \operatorname{diag}(B,A) \qquad B\ m \times m \qquad A\ n \times n \tag{17-4}$$

and all eigenvalues of B have simple elementary divisors and zero real parts and no eigenvalues of A have zero real parts. According to our convention in Chap. 13, the system $\dot{z} = Dz$, with D as above, is *critical*.

If the vectors z, q are partitioned as $z = \operatorname{col}(z^1,z^2)$, $q = \operatorname{col}(f,g)$, where z^1, f are m vectors, then the almost periodic transformation

$$z^1 = e^{Bt}x \qquad z^2 = y \tag{17-5}$$

yields the equivalent system

$$\begin{aligned} \dot{x} &= \epsilon e^{-Bt}f(t,e^{Bt}x,y,\epsilon) \\ \dot{y} &= Ay + \epsilon g(t,e^{Bt}x,y,\epsilon) \end{aligned} \tag{17-6}$$

This system (17-6) is clearly a special case of system (17-1), and we can state the following consequence of Theorem 17-1.

Theorem 17-3. If system (17-3), (17-4) satisfies the conditions enumerated above, if z, q are partitioned as $z = \operatorname{col}(x,y)$, $q = \operatorname{col}(f,g)$, where x, f are m vectors and y, g are n vectors, then, for ϵ sufficiently small, there exists an almost periodic solution of (17-3) if there is a constant m vector x^0 such that

$$0 = F(x^0) \stackrel{\text{def}}{\equiv} \lim_{T\to\infty} T^{-1} \int_0^T e^{-Bt} f(t, e^{Bt}x^0, 0, 0)\, dt \tag{17-7}$$

and the eigenvalues of the matrix $\partial F(x^0)/\partial x$ have nonzero real parts. Furthermore, this almost periodic solution $z^*(t,\epsilon)$ has the same basic frequencies as the almost periodic function col $[e^{-Bt}f(t,e^{Bt}x,y,\epsilon), g(t,e^{Bt}x,y,\epsilon)]$, satisfies $z^*(t,0) = \text{col}\,(e^{Bt}x^0,0)$, and for $\epsilon > 0$ the stability properties of z^* are locally the same as the stability properties of the solution $x = y = 0$ of $\dot{x} = [\partial F(x^0)/\partial x]x$, $\dot{y} = Ay$.

A result analogous to Theorem 17-3 can also be proved for the more general system

$$\dot{z} = Dz + \epsilon q(t,z,\epsilon) + h(t) \tag{17-8}$$

where D, q, ϵ are as in system (17-3), (17-4) and $h(t)$ is almost periodic in t. The transformation (17-5) transforms (17-8) into a system of the form (17-6) except with the first equation perturbed by the vector $e^{-Bt}h^1(t)$, $h = \text{col}\,(h^1,h^2)$, h^1 an m vector. Consequently, if

$$\int_0^t e^{-B\tau}h^1(\tau)\, d\tau$$

is a bounded function of t and one lets $u(t)$ be any primitive of $e^{-Bt}h^1(t)$, then the transformation $x \to u + x$, $y \to y$ reduces this apparently more complicated case to the previous one.

Example 17-1. *Forced van der Pol Equation.* Consider the equation

$$\begin{aligned} \dot{z}_1 &= z_2 \\ \dot{z}_2 &= -z_1 + \epsilon(1 - z_1^2)z_2 + A \sin \omega_1 t + B \sin \omega_2 t \end{aligned} \tag{17-9}$$

where $\epsilon > 0$, $\omega_1 > 0$, $\omega_2 > 0$, A, B are constants and

$$m + m_1\omega_1 + m_2\omega_2 \neq 0$$

for all integers m, m_1, m_2 with $|m| + |m_1| + |m_2| \leqq 4$. In the previous section, we have shown that, for *any* fixed $\epsilon > 0$, system (17-9) has a stable cylinder of solutions if the constants A, B are sufficiently small. With the results obtained in this section we wish to find some more properties of this equation for ϵ small and A, B arbitrary. For $\epsilon = 0$, the general solution of (17-9) is

$$\begin{aligned} z_1 &= x_1 \cos t + x_2 \sin t + A(1 - \omega_1^2)^{-1} \sin \omega_1 t \\ &\qquad + B(1 - \omega_2^2)^{-1} \sin \omega_2 t \\ z_2 &= -x_1 \sin t + x_2 \cos t + A\omega_1(1 - \omega_1^2)^{-1} \cos \omega_1 t \\ &\qquad + B\omega_2(1 - \omega_2^2)^{-1} \cos \omega_2 t \end{aligned} \tag{17-10}$$

where x_1, x_2 are arbitrary constants. To discuss the existence of almost periodic solutions of (17-9), we make the transformation (17-10) taking x_1, x_2 as new variables. The new differential equations for x_1, x_2 are

$$\begin{aligned} \dot{x}_1 &= -\epsilon(1 - z_1^2)z_2 \sin t \\ \dot{x}_2 &= \epsilon(1 - z_1^2)z_2 \cos t \end{aligned} \tag{17-11}$$

where z_1, z_2 are given by (17-10). If $x = \operatorname{col}(x_1,x_2)$, this system is a special case of the system (17-1) with the y variable absent.

The averaged equations associated with (17-11) are

$$\begin{aligned} 8\dot{x}_1 &= \epsilon x_1[2(2 - g(A,B,\omega_1,\omega_2)) - (x_1^2 + x_2^2)] \\ 8\dot{x}_2 &= \epsilon x_2[2(2 - g(A,B,\omega_1,\omega_2)) - (x_1^2 + x_2^2)] \\ g(A,B,\omega_1,\omega_2) &= A^2(1 - \omega_1^2)^{-2} + B^2(1 - \omega_2^2)^{-2} \end{aligned} \tag{17-12}$$

The equations (17-12) have the constant solution $x_1 = x_2 = 0$ which is asymptotically stable if $g(A,B,\omega_1,\omega_2) > 2$ and unstable if $g(A,B,\omega_1,\omega_2) < 2$. Therefore, from Theorem 17-1, for $g(A,B,\omega_1,\omega_2) \neq 2$, there exists a unique almost periodic solution $z(t,\epsilon)$ of (17-9) for

$$0 < \epsilon \leqq \epsilon_0 = \epsilon_0(A,B,\omega_1,\omega_2)$$

such that

$$\begin{aligned} z_1(t,0) &= A(1 - \omega_1^2)^{-1} \sin \omega_1 t + B(1 - \omega_2^2)^{-1} \sin \omega_2 t \\ z_2(t,0) &= \dot{z}_1(t,0) \end{aligned} \tag{17-13}$$

and this solution is stable or unstable according as $g(A,B,\omega_1,\omega_2)$ is >2 or <2.

In the unstable case, $g(A,B,\omega_1,\omega_2) < 2$, system (17-12) also has a family of equilibrium points given by $x_1 =$ constant, $x_2 =$ constant,

$$x_1^2 + x_2^2 = 2(2 - g(A,B,\omega_1,\omega_2))$$

In this case, one of the roots of the linear variational equation is zero and Theorem 17-1 does not apply. We return to this example again in the next chapter to obtain more information about the behavior of the solutions for ϵ sufficiently small and $g(A,B,\omega_1,\omega_2) < 2$. However, from what we have learned in Chap. 16 we know what to expect. In fact, two of the simplest ways to make $g(A,B,\omega_1,\omega_2) < 2$ are to require that either A and B are sufficiently small or ω_1 and ω_2 are sufficiently large. In Chap. 16, we have already remarked that these cases lead to a stable cylinder of solutions.

We return to this example in the next chapter to understand more about the case $g(A,B,\omega_1,\omega_2) < 2$.

Let us now consider the system

$$\dot{z} = Dz + \epsilon\Phi(t)z \tag{17-14}$$

where z is an $(n + m)$ vector, D is a constant matrix, and $\Phi(t)$ is almost periodic in t.

Theorem 17-4. Suppose $D = \operatorname{diag}(B,A)$ where B is $m \times m$, A is $n \times n$, the eigenvalues of B have simple elementary divisors and zero real parts, and all the eigenvalues of A have negative real parts. If the almost periodic matrix $\Phi(t)$ is partitioned as $\Phi(t) = (\Phi_{jk}(t))$, $j, k = 1, 2$, where $\Phi_{11}(t)$, $\Phi_{22}(t)$ are $m \times m$, $n \times n$ matrices, respectively and if all the eigenvalues of the matrix

$$\lim_{T\to\infty} T^{-1} \int_0^T e^{-Bt}\Phi_{11}(t)e^{Bt}\,dt \tag{17-15}$$

have negative real parts, then, for $\epsilon > 0$ sufficiently small, all solutions of system (17-14) approach zero exponentially as $t \to \infty$.

Proof. The transformation (17-5) yields the equivalent system

$$\begin{aligned} \dot{x} &= \epsilon e^{-Bt}\Phi_{11}(t)e^{Bt}x + \epsilon e^{-Bt}\Phi_{12}(t)y \\ \dot{y} &= Ay + \epsilon\Phi_{21}e^{Bt}x + \epsilon\Phi_{22}(t)y \end{aligned}$$

This system has the almost periodic solution $x = y = 0$ and is clearly a special case of system (17-1). The result then follows immediately from Theorem 17-1 and the fact that (17-14) is a linear equation.

In case Φ in (17-14) is a periodic system, then the eigenvalues of the matrix (17-15) yield the first approximations to the characteristic exponents of (17-14) which for $\epsilon = 0$ coincide with the eigenvalues of B. In fact, if $\lambda_1, \ldots, \lambda_n$ are the eigenvalues of B and $\sigma_1, \ldots, \sigma_n$ are the eigenvalues of (17-15), then the first-order approximation to the characteristic exponents close to $\lambda_1, \ldots, \lambda_n$ are given by $\lambda_1 + \epsilon\sigma_1, \ldots, \lambda_n + \epsilon\sigma_n$. The matrix (17-15) is very easy to calculate and in many cases decides the question of stability. This same result for periodic systems was stated in Chap. 8.

18: Integral Manifolds—Critical Case

In this chapter we state and apply some results which include those of the previous chapter.

Consider the system of equations

$$
\begin{aligned}
\dot{\theta} &= d + \epsilon\Theta(t,\theta,x,y,\epsilon) \\
\dot{x} &= \epsilon X(t,\theta,x,y,\epsilon) \\
\dot{y} &= Ay + \epsilon Y(t,\theta,x,y,\epsilon)
\end{aligned}
\tag{18-1}
$$

where ϵ is a real parameter, θ, d, Θ are k vectors, x, X are m vectors, y, Y are n vectors, A is an $n \times n$ constant matrix whose eigenvalues have nonzero real parts, $d = \text{col}\,(1, 1, \ldots, 1)$, Θ, X, Y are bounded and uniformly continuous together with their first partial derivatives with respect to t, θ, x, y and their second partial derivatives with respect to x, y for $-\infty < t < \infty$, $\theta \in E^k$, $x \in U$, $y \in V$, $0 \leqq \epsilon \leqq \epsilon_0$, $\epsilon_0 > 0$, where U, V are open sets in E^m, E^n, respectively. Furthermore, the functions Θ, X, Y are multiply periodic in θ of vector period ω and, for each fixed ϵ, $0 \leqq \epsilon \leqq \epsilon_0$, are almost periodic in t uniformly with respect to θ, x, y, for $\theta \in E^k$, $x \in U$, $y \in V$.

Theorem 18-1. Suppose system (18-1) satisfies the conditions enumerated above and that the mean value

$$\lim_{T\to\infty} T^{-1} \int_0^T X(t + \tau, \theta + \tau, x, y, \epsilon)\, d\tau$$

$$\theta + \tau = \text{col}\,(\theta_1 + \tau, \ldots, \theta_k + \tau)$$

is independent of t, θ and designate this function by $X_0(x,y,\epsilon)$.

If there exists a constant m vector x^0 such that $X(x^0,0,0) = 0$ and the eigenvalues of the matrix $\partial X\,(x^0,0,0)/\partial x$ have nonzero real parts, then it is possible to find a positive constant ϵ_1 and vector functions

$f(t,\theta,\epsilon)$, $g(t,\theta,\epsilon)$ of dimensions m, n, respectively, such that $f(t,\theta,\epsilon)$, $g(t,\theta,\epsilon)$ are continuous in t, θ, ϵ, $-\infty < t < \infty$, $\theta \in E^k$, $0 \leqq \epsilon \leqq \epsilon_1$, multiply periodic in θ of vector period ω and for each fixed ϵ, $0 \leqq \epsilon \leqq \epsilon_1$ almost periodic in t uniformly with respect to $\theta \in E^k$, $f(t,\theta,0) = x^0$, $g(t,\theta,0) = 0$ for all $t \in E$, $\theta \in E^k$, such that $x = f(t,\theta,\epsilon)$, $y = g(t,\theta,\epsilon)$ is an integral manifold of system (18-1). The stability properties of this integral manifold are locally the same as the stability properties of the solution $x = 0$, $y = 0$ of the system

$$\dot{x} = \frac{\partial X\,(x^0,0,0)}{\partial x}\,x \qquad \dot{y} = Ay$$

The proof of this theorem is very similar to the proof of Theorem 17-1. Also, in a manner similar to the proof of Theorem 17-2 one proves the following theorem.

Theorem 18-2. If system (18-1) satisfies all the conditions enumerated above, if $X_0(x,y,\epsilon)$ is defined as in Theorem 18-1 and there exists a periodic solution $x^0(\epsilon\omega t)$, $x^0(s + 2\pi) = x^0(s)$, $x^0(s) \in U$, $0 \leqq s \leqq 2\pi$, of the equation

$$\dot{x} = \epsilon X_0(x,0,0)$$

such that the linear variational equation has $(m - 1)$ characteristic exponents with nonzero real parts, then there exists a positive constant ϵ_1 such that, for every ϵ, $0 < \epsilon \leqq \epsilon_1$, the following conclusions hold:

1. Equation (18-1) has an integral manifold S which has a parametric representation of the form

$$x = f(t,\theta,\varphi,\epsilon) \qquad y = g(t,\theta,\varphi,\epsilon)$$
$$-\infty < t < \infty \qquad \theta \in E^k \qquad 0 \leqq \varphi \leqq 2\pi$$

where f, g are continuous in t, θ, φ, ϵ, multiply periodic in θ of vector period ω, periodic in φ of period 2π, and for each fixed ϵ, $0 < \epsilon \leqq \epsilon_1$, almost periodic in t uniformly with respect to θ, φ for $\theta \in E^k$, $0 \leq \varphi \leq 2\pi$. Furthermore,

$$\|f(t,\theta,\varphi,\epsilon) - x^0(\varphi)\| \to 0 \qquad \|g(t,\theta,\varphi,\epsilon)\| \to 0 \qquad \text{as } \epsilon \to 0$$

uniformly with respect to $t \in E$, $\theta \in E^k$, $\varphi \in E$.

2. If the cylinder $x = x^0(\varphi)$, $y = 0$, $0 \leqq \varphi \leqq 2\pi$, $-\infty < t < \infty$, which is an integral manifold of

$$\dot{x} = \epsilon X(x,0,0) \qquad \dot{y} = Ay$$

is stable, unstable, or conditionally stable with respect to a manifold of dimension s, then the integral manifold S of (18-1) is stable, unstable, or conditionally stable with respect to a manifold of dimension s.

Consider the system of second-order equations

$$\begin{aligned} \ddot{z}_j + \mu_j^2 z_j &= \epsilon Z_j(t, z_1, \ldots, z_m, \dot{z}_1, \ldots, \dot{z}_m) \\ j &= 1, 2, \ldots, m \end{aligned} \tag{18-2}$$

where each $\mu_j > 0$. It is a very simple matter to transform equation (18-2) into the form (18-1). In fact, if

$$z_j = x_j \sin \mu_j\theta_j \qquad \dot{z}_j = x_j\mu_j \cos \mu_j\theta_j \qquad j = 1, 2, \ldots, m \tag{18-3}$$

then

$$\begin{aligned} \dot{\theta}_j &= 1 + \frac{\epsilon}{\mu_j^2 x_j} Z_j(t, x \sin \mu\theta, x\mu \cos \mu\theta) \sin \mu_j\theta_j \\ \dot{x}_j &= \frac{\epsilon}{\mu_j} Z_j(t, x \sin \mu\theta, x\mu \cos \mu\theta) \cos \mu_j\theta_j \\ j &= 1, 2, \ldots, m \end{aligned} \tag{18-4}$$

where $x \sin \mu\theta$, $x\mu \cos \mu\theta$ is a shorthand notation for

$$\begin{aligned} x \sin \mu\theta &= (x_1 \sin \mu_1\theta_1, \ldots, x_m \sin \mu_m\theta_m) \\ x\mu \cos \mu\theta &= (x_1\mu_1 \cos \mu_1\theta_1, \ldots, x_m\mu_m \cos \mu_m\theta_m) \end{aligned}$$

Theorems 18-1 and 18-2 can now be applied to system (18-4) with $\theta = \text{col}(\theta_1, \ldots, \theta_m)$, $x = \text{col}(x_1, \ldots, x_m)$ and the variable y absent. In particular, for $m = 1$ and Z_1 independent of t, that is, a single second-order autonomous equation, one obtains the very simple criterion for the existence of a periodic solution. If

$$F(x) = \int_0^{2T/\mu} Z(x \sin \mu\theta, x\mu \cos \mu\theta) \cos \mu\theta \, d\theta \tag{18-5}$$

and there exists an x^0 such that $F(x^0) = 0$, $F'(x^0) \neq 0$, then there exists, for ϵ sufficiently small, a periodic solution of the second-order equation

$$\ddot{z} + \mu^2 z = \epsilon Z(z,\dot{z}) \tag{18-6}$$

which for $\epsilon = 0$ coincides with the function $z = x^0 \sin \mu t$. More precisely there exists a function $f(\theta,\epsilon)$ periodic in θ of period $2\pi/\mu$, $f(\theta,0) = x^0$, such that the cylinder

$$z = f(\theta,\epsilon) \sin \mu\theta \qquad \dot{z} = \mu f(\theta,\epsilon) \cos \mu\theta \tag{18-7}$$

is an integral manifold of (18-6). Furthermore, the period of the periodic solutions on this manifold are obtained by solving the equation

$$\dot{\theta} = 1 - \frac{\epsilon}{\mu^2 f(\theta,\epsilon)} Z(f(\theta,\epsilon) \sin \mu\theta, f(\theta,\epsilon)\mu \cos \mu\theta) \sin \mu\theta$$

for $\theta = \theta(t)$ and finding the inverse function $t = t(\theta)$. If $\epsilon F'(x^0) < 0$, this manifold is stable, and if $\epsilon F'(x^0) > 0$, this manifold is unstable. These results agree with the first approximations obtained in Chap. 10 when we discussed periodic solutions by another method.

The advantage of the method of averaging, however, lies in the fact that we can discuss more complicated phenomena. This is illustrated by the following examples.

Example 18-1. *Forced van der Pol Equation.* In Chap. 17, we discussed some aspects of system (17-9); namely, the existence of almost periodic solutions was obtained from the results of Chap. 17. Also, the fact that, for fixed ϵ, there should be a two-dimensional integral manifold if either A, B are sufficiently small or ω_1, ω_2 are sufficiently large followed from the results in Chap. 15. Furthermore, we noticed that, if the constant $g(A,B,\omega_1,\omega_2)$ in (18-12) was <2, then the almost periodic solution $z(t,\epsilon)$ of (17-9) with $z(t,0)$ given by (17-13) was unstable, and if $g(A,B,\omega_1,\omega_2)$ was >2, then $z(t,\epsilon)$ was stable. We wish now to discuss the case where $g(A,B,\omega_1,\omega_2) < 2$ in more detail.

For $\epsilon = 0$, the general solution of (17-9) can be written as

$$\begin{aligned} z_1 &= x \sin \theta + A(1 - \omega_1^2)^{-1} \sin \omega_1 t + B(1 - \omega_2^2)^{-1} \sin \omega_2 t \\ z_2 &= x \cos \theta + A\omega_1(1 - \omega_1^2)^{-1} \cos \omega_1 t + B\omega_2(1 - \omega_2^2)^{-1} \cos \omega_2 t \end{aligned} \tag{18-8}$$

where $\theta = t + \varphi$, and φ, x are arbitrary constants. Now, treating x, θ as new coordinates, we obtain

$$\begin{aligned} \dot{x} &= \epsilon(1 - z_1^2)z_2 \cos \theta \\ \dot{\theta} &= 1 - (\epsilon/x)(1 - z_1^2)z_2 \sin \theta \end{aligned} \tag{18-9}$$

where z_1, z_2 are given in (18-8). This system is of the same type as the one considered in the above theorems.

Since $m + m_1\omega_1 + m_2\omega_2 \neq 0$ for all integers m, m_1, m_2 with $|m| + |m_1| + |m_2| \leqq 4$, the averaged equations associated with (18-9) are

$$\begin{aligned} 8\dot{x} &= \epsilon x[2(2 - g(A,B,\omega_1,\omega_2)) - x^2] \\ \dot{\theta} &= 1 \end{aligned} \tag{18-10}$$

where $g(A,B,\omega_1,\omega_2)$ is the same function as given in (17-12). System (18-10) always has the solution $x = 0$, $\theta = t + \varphi$, where φ is an arbitrary constant. This solution will correspond to the almost periodic solution obtained before.

Now, if $g(A,B,\omega_1,\omega_2) < 2$, then system (18-10) also has the stable family of solutions $(x^0)^2 = 2[2 - g(A,B,\omega_1,\omega_2)]$, $\theta = t + \varphi$, where φ is an arbitrary constant. Therefore, Theorem 18-1 asserts the existence of a function $f(t,\theta,\epsilon)$, $0 \leqq \epsilon \leqq \epsilon_0$, $f(t,\theta,0) = x^0$,

$$f(t,\theta,\epsilon) = f(t, \theta + 2\pi, \epsilon)$$

$f(t,\theta,\epsilon)$ almost periodic in t with frequencies ω_1, ω_2, such that there is a one-parameter family of solutions of (18-9) given by

$$\begin{aligned} x &= f(t,\theta,\epsilon) \\ \dot{\theta} &= 1 - (\epsilon/x)(1 - z_1^2)z_2 \sin \theta \qquad \theta(0) = \theta_0 \end{aligned} \tag{18-11}$$

where θ_0 is an arbitrary real number and z_1, z_2 are given by (18-8) with x replaced by $f(t,\theta,\epsilon)$.

In terms of the original variable z, we can say that there is a family of solutions of (17-9) given by

$$\begin{aligned} z_1 &= f(t,\theta,\epsilon) \sin \theta + A(1 - \omega_1^2)^{-1} \sin \omega_1 t \\ &\qquad + B(1 - \omega_2^2)^{-1} \sin \omega_2 t \\ z_2 &= f(t,\theta,\epsilon) \cos \theta + A\omega_1(1 - \omega_1^2)^{-1} \cos \omega_1 t \\ &\qquad + B\omega_2(1 - \omega_2^2)^{-1} \cos \omega_2 t \end{aligned} \tag{18-12}$$

where θ satisfies the previous equations (18-11).

Furthermore, if we let M be the set of points in (x,t) space defined by (18-12) with θ varying from $-\infty$ to $+\infty$ (M is a generalized cylinder), and if $Z(t)$ is a solution of (17-9) with initial value close enough to M, then $x(t) \to M$ as $t \to \infty$. Thus, from the practical point of view, the solutions defined by (18-12) possess stability.

Let us try to understand a little better the geometry of the solution curves for this example. For any value of the parameter $g(A,B,\omega_1,\omega_2) \neq 2$, there always exists an almost periodic solution of system (17-9) for which ϵ small is very close to the solution (17-13) of the linear nonhomogeneous equation

$$\begin{aligned} \dot{z}_1 &= z_2 \\ \dot{z}_2 &= -z_1 + A \sin \omega_1 t + B \sin \omega_2 t \end{aligned} \tag{18-13}$$

Furthermore, this solution is stable if $g(A,B,\omega_1,\omega_2) > 2$ and unstable if $g(A,B,\omega_1,\omega_2) < 2$. For $g(A,B,\omega_1,\omega_2)$ very close to zero, there is an integral manifold of (17-9) which is no longer close to the almost periodic solution of (18-13) even for ϵ small and, in fact, solutions on this integral manifold can have an amplitude which differs by approximately two units from the solution of (18-13). As $g(A,B,\omega_1,\omega_2) \to 2$ the integral manifold and the almost periodic solution come closer and closer together.

If we let

$$x_{01}(t) = A(1 - \omega_1^2)^{-1} \sin \omega_1 t + B(1 - \omega_2^2)^{-1} \sin \omega_2 t$$
$$x_{02}(t) = \dot{x}_{01}(t)$$

then, except for the translation $x_{01}(t)$, $x_{02}(t)$, the first approximation ($\epsilon = 0$) to the above integral manifolds of system (17-9) are shown in Fig. 18-1.

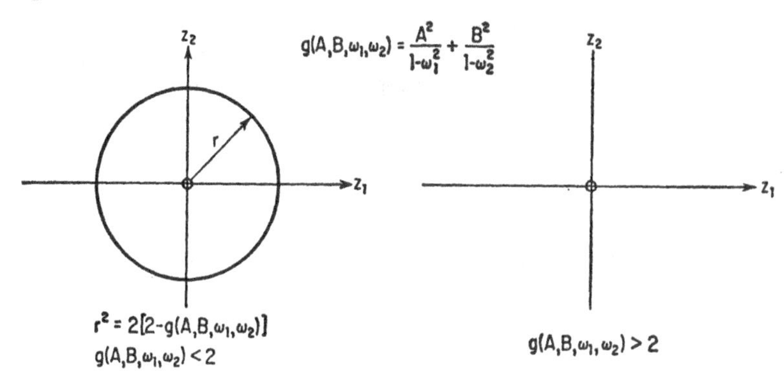

Fig. 18-1

Example 18-2. *Coupled van der Pol Equations.** Consider the system of second-order equations

$$\begin{aligned} \ddot{z}_1 + \mu_1^2 z_1 &= \epsilon(1 - z_1^2 - az_2^2 - bz_1^2 z_2^2)\dot{z}_1 \\ \ddot{z}_2 + \mu_2^2 z_2 &= \epsilon(1 - \alpha z_1^2 - z_2^2)\dot{z}_2 \end{aligned} \tag{18-14}$$

where z_1, z_2 are scalars, $\epsilon > 0$, $a > 0$, $\alpha > 0$ and b are parameters. Furthermore, suppose that $\mu_1 > 0$, $\mu_2 > 0$ are such that $\mu_j \not\equiv 0 \pmod{\mu_k}$, $j \neq k$, and $k\mu_2 + l\mu_1 \neq 0$ for all integers k, l such that $|k| + |l| \leq 3$.

System (18-14) always has the solution $z_1 = z_2 = 0$ for all t, and this solution is unstable.

By using the procedure outlined in Chap. 10, it is not difficult to see that, for ϵ sufficiently small, there are always two periodic solutions of system (18-14):

(P_1) $\quad z_{11}(t,\epsilon)$, $z_{21}(t,\epsilon)$, of period $2\pi/\tau_1(\epsilon)$, $\tau_1(0) = \mu_1$
$\quad z_{11}(t,0) = 2 \cos \mu_1 t \qquad z_{21}(t,0) = 0$

(P_2) $\quad z_{12}(t,\epsilon)$, $z_{22}(t,\epsilon)$, of period $2\pi/\tau_2(\epsilon)$, $\tau_2(0) = \mu_2$
$\quad z_{12}(t,0) = 0 \qquad z_{22}(t,0) = 2 \cos \mu_2 t$

By applying the stability Theorem 10-1, the solution (P_1) is asymptotically orbitally stable if $\alpha > \frac{1}{2}$ and unstable if $\alpha < \frac{1}{2}$. The

* The author is indebted to Gene Lefferts for the computations in this example.

solution (P_2) is asymptotically orbitally stable if $a > \frac{1}{2}$ and unstable if $a < \frac{1}{2}$. By analyzing in a little more detail the characteristic exponents of the linear variational equation associated with these periodic solutions, one observes that, when either solution is unstable there are one zero characteristic exponent, one negative characteristic exponent, and two characteristic exponents which have a positive real part and nonzero imaginary part.

Let us now analyze system (18-14) for other types of solutions by the methods of this section. More specifically, we wish to investigate the possibility of the existence of integral manifolds for system (18-14) which are more complicated than the cylinders generated by the periodic solutions. The transformation

$$z_j = x_j \sin \mu_j\theta_j \qquad \dot{z}_j = x_j\mu_j \cos \mu_j\theta_j \qquad j = 1, 2 \tag{18-15}$$

yields the equivalent system

$$\begin{aligned}
\dot{\theta}_1 &= 1 + \frac{\epsilon}{\mu_1^2 x_1}(1 - z_1^2 - az_2^2 - bz_1^2z_2^2)\dot{z}_1 \sin \mu_1\theta_1 \\
\dot{\theta}_2 &= 1 + \frac{\epsilon}{\mu_2^2 x_2}(1 - \alpha z_1^2 - z_2^2)\dot{z}_2 \sin \mu_2\theta_2 \\
\dot{x}_1 &= \frac{\epsilon}{\mu_1}(1 - z_1^2 - az_2^2 - bz_1^2z_2^2)\dot{z}_1 \cos \mu_1\theta_1 \\
\dot{x}_2 &= \frac{\epsilon}{\mu_2}(1 - \alpha z_1^2 - z_2^2)\dot{z}_2 \cos \mu_2\theta_2
\end{aligned} \tag{18-16}$$

where z_j, $\dot{z}_j$ are given in terms of x_j, θ_j by (18-15). From the assumptions above on the μ_j, if we average the right-hand sides of (18-16) with respect to θ_1, θ_2 according to the formula in Theorem (18-1), we obtain for the averaged equations

$$\begin{aligned}
&\dot{\theta}_1 = 1 \qquad \dot{\theta}_2 = 1 \\
&\dot{x}_1 = \frac{\epsilon x_1}{2}\left(1 - \frac{x_1^2}{4} - \frac{ax_2^2}{2} - \frac{bx_1^2x_2^2}{4}\right) \\
&\dot{x}_2 = \frac{\epsilon x_2}{2}\left(1 - \frac{\alpha x_1^2}{2} - \frac{x_2^2}{4}\right)
\end{aligned} \tag{18-17}$$

From Theorem 18-1, we know that if x_{10}, x_{20} is an equilibrium point of the equations in x_1, x_2 in (18-17) and the corresponding coefficient matrix of the linear variational equations with respect to x_{10}, x_{20} have nonzero real parts, then, for ϵ sufficiently small, there exist functions $f(\theta_1,\theta_2,\epsilon)$, $g(\theta_1,\theta_2,\epsilon)$ which are periodic in θ_1, θ_2 of periods $2\pi/\mu_1$, $2\pi/\mu_2$ respectively, and $f(\theta_1,\theta_2,0) = x_{10}$, $f(\theta_1,\theta_2,0) = x_{20}$. From (18-15) this implies there is a manifold of solutions of (18-14) of the form

$$\begin{aligned} z_1 &= f(\theta_1,\theta_2,\epsilon) \sin \mu_1\theta_1 & \dot{z}_1 &= \mu_1 f(\theta_1,\theta_2,\epsilon) \cos \mu_1\theta_1 \\ z_2 &= g(\theta_1,\theta_2,\epsilon) \sin \mu_2\theta_2 & \dot{z}_2 &= \mu_2 g(\theta_1,\theta_2,\epsilon) \cos \mu_2\theta_2 \end{aligned} \tag{18-18}$$

The stability properties of this manifold are determined by the eigenvalues of the linear variational equation of (18-17). Of course, the dimension of this manifold does not have to be 3 in $(z_1,\dot{z}_1,z_2,\dot{z}_2,t)$ space [of dimension 2 in $(z_1,\dot{z}_1,z_2,\dot{z}_2)$ space] since both parameters θ_1, θ_2 need not appear in f, g. In fact, (18-17) has the equilibrium point $x_1 = x_2 = 0$ which corresponds to the zero solution of (18-14) and f, g are independent of θ_1, θ_2. Equations (18-17) also have the solution $x_1 = 2$, $x_2 = 0$ which corresponds to the periodic solution (P_1) above and f, g do not contain θ_2. There is also the solution $x_1 = 0$, $x_2 = 2$, which corresponds to the periodic solution (P_2) above, and f, g do not contain θ_1.

Now, we ask the following question: Is it possible to have an equilibrium point $x_{10} \neq 0$, $x_{20} \neq 0$ of (18-17) with the coefficient matrix of the linear variational equation having eigenvalues with nonzero real parts? If such an equilibrium point exists, then there will be an integral manifold of (18-14) of the form (18-18) where f, g depend upon both θ_1, θ_2 and thus this will be an integral manifold of dimension 2 in $(z_1,\dot{z}_1,z_2,\dot{z}_2)$ space. Since f, g are multiply periodic in θ_1, θ_2 this manifold is a torus (doughnut) and the behavior of the solutions on a torus is well known (see Coddington and Levinson [1]). The solutions may be almost periodic, periodic, or approach periodic solutions.

We consider the two cases where $b = 0$ and $b \neq 0$. If $b = 0$, it is very easy to find the equilibrium points of (18-17) and, in fact, one observes that if either (1) $a < \frac{1}{2}$, $\alpha < \frac{1}{2}$, or, if (2) $a > \frac{1}{2}$, $\alpha > \frac{1}{2}$, then there are equilibrium points $x_{10} > 0$, $x_{20} > 0$ of (18-17) given by

$$x_{10} = 2\sqrt{\frac{1-2a}{1-4a\alpha}} \qquad x_{20} = 2\sqrt{\frac{1-2\alpha}{1-4a\alpha}} \tag{18-19}$$

Furthermore, in case 1 both the eigenvalues of the coefficient matrix of the linear variational equation associated with x_{10}, x_{20} are negative and, in case 2, one is positive and one is negative. In either case, the conditions of Theorem 18-1 are satisfied and we obtain an integral manifold of (18-14) of dimension 2 in $(z_1,\dot{z}_1,z_2,\dot{z}_2)$ space. In case 1 this manifold is asymptotically stable and in case 2 it is unstable. These results are summarized in Fig. 18-2, where we have designated by (P_1), (P_2) the periodic solutions of (18-14) mentioned above and by (M^2) the integral manifold generated by x_{10}, x_{20} in (18-19).

If $b \neq 0$, then it becomes rather difficult to find exactly the equilibrium points of (18-17). In general, the behavior will be the same as that depicted in Fig. 18-2. However, with $b \neq 0$, there are also situations where one can obtain two integral manifolds of the type M^2 above, and also when one obtains only one solution of the type M^2 the type of stability may change by a variation in b.

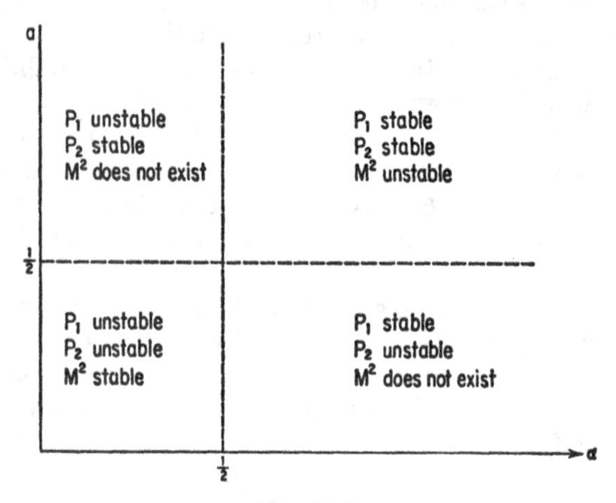

Fig. 18-2

For some values of a, α, b, the types of behavior that can occur are depicted in Table 18-1. In this table, we say a manifold of type M^2 above is a saddle (stable node) (stable focus) if the eigenvalues of the coefficient matrix of the linear variational equation associated with an equilibrium point (x_{10}, x_{20}) of (18-17) have one root positive and

Table 18-1

a	α	b	P_1	P_2	M_1^2	M_2^2
0.25	0.25	0	Unstable	Unstable	Stable node	
		1	Unstable	Unstable	Stable node	
		−1	Unstable	Unstable	Stable focus	
0.8	0.8	0	Stable	Stable	Saddle	
		1	Stable	Stable	Saddle	
		−1	Stable	Stable	Saddle	
0.25	0.8	0	Stable	Unstable		
		1	Stable	Unstable	Saddle	Stable node

one negative (both negative) (real parts negative and nonzero imaginary part). The solutions of type M^2 (since there are sometimes two) are designated by M_1^2, M_2^2. If nothing occurs in the column M_2^2, this means there is only one solution of type M^2.

As one can see, coupled van der Pol equations of the type (18-14) can have many different types of oscillatory behavior, which seem to be of interest. Of course, an understanding of the behavior of the trajectories of system (18-14) in four-dimensional space and the nature of the transition from the existence of an integral manifold of type M^2 to the nonexistence of such a manifold is lacking. Behavior similar to that described above has been observed by Plaat [1].

APPENDIX:

Principle of Contraction Mappings

A *metric space* is a set X, whose elements are called points, and a distance; that is, a single-valued nonnegative real function $\rho(x,y)$ defined for arbitrary x, y in X and satisfying the following conditions:

1. $\rho(x,y) = 0$ if and only if $x = y$
2. $\rho(x,y) = \rho(y,x)$
3. $\rho(x,y) + \rho(y,z) \geqq \rho(x,z)$

The metric space will be designated by X without mentioning the distance function ρ. A simple example of a metric space is the space E^n introduced in Chap. 2 with $\rho(x,y) = \|x - y\|$, and $\|x\|$ defined on page 15.

A mapping f of a metric space X into itself is said to be *continuous at a point* x_0 in X if for any $\epsilon > 0$ there is a $\delta > 0$ such that $\rho(x,x_0) < \delta$ implies $\rho[f(x), f(x_0)] < \epsilon$.

A sequence $\{x_n\}$ of points in a metric space X is called a *Cauchy sequence* if for arbitrary $\epsilon > 0$ there exists an N_ϵ such that $\rho(x_n,x_m) < \epsilon$ for all $n \geqq N_\epsilon$, $m \geqq N_\epsilon$. A sequence $\{x_n\}$ of points in a metric space X *converges* to a point x if for any $\epsilon > 0$ there is an N_ϵ such that $\rho(x_n,x) < \epsilon$ for all $n \geqq N_\epsilon$. A metric space X is called *complete* if every Cauchy sequence converges to an element in X. It is a classical result that the space E^n introduced in Chap. 2 is complete.

A mapping f of a metric space X into itself is called a *contraction mapping* if there exists a number $K < 1$ such that for all x, y in X, $\rho[f(x), f(y)] \leqq K\rho(x,y)$. A contraction mapping is necessarily continuous. For example, if f is a function of R^1 into itself with

$$\rho(x,y) = |x - y|$$

then f is a contraction mapping if

$$|f(x) - f(y)| \leqq K|x - y| \qquad K < 1$$

that is, if f is Lipschitzian with Lipschitz constant $K < 1$.

Theorem A-1. (Principle of Contraction Mappings.) Every contraction mapping defined in a complete metric space X has one and only one fixed point in X; that is, there is one and only one point in X such that $x = f(x)$.

Proof. Let x_0 be an arbitrary point in X and consider the sequence of successive approximations

$$x_{n+1} = f(x_n) \qquad n = 0, 1, 2, \ldots \tag{A-1}$$

We now show that the sequence $\{x_n\}$ is a Cauchy sequence. In fact, if $m \geqq n$, there is a constant $K < 1$ such that

$$\begin{aligned}
\rho(x_n,x_m) &= \rho[f(x_{n-1}), f(x_{m-1})] \leqq K\rho(x_{n-1},x_{m-1}) \\
&\leqq K^n\rho(x_0,x_{m-n}) \\
&\leqq K^n[\rho(x_0,x_1) + \rho(x_1,x_2) + \cdots + \rho(x_{m-n-1}, x_{m-n})] \\
&\leqq K^n\rho(x_0,x_1)(1 + K + \cdots + K^{m-n-1}) \\
&\leqq \frac{K^n}{1-K}\rho(x_0,x_1)
\end{aligned}$$

Since $K < 1$, this quantity is arbitrarily small for sufficiently large n, and $\{x_n\}$ is a Cauchy sequence. Since X is complete, there exists an x in X such that $\lim_{n\to\infty} x_n = x$. Since the mapping f is continuous

$$f(x) = f(\lim_{n\to\infty} x_n) = \lim_{n\to\infty} f(x_n) = \lim_{n\to\infty} x_{n+1} = x$$

and the existence of a fixed point is proved.

To prove uniqueness, suppose there exist x, y in X such that $x = f(x)$, $y = f(y)$. Then

$$\rho(x,y) = \rho[f(x), f(y)] \leqq K\rho(x,y)$$

and, since $K < 1$, $\rho(x,y) = 0$ which implies $x = y$.

From the above theorem, we see that one can find a fixed point of a mapping f in X if one shows that, for any x in X, $f(x)$ is in X and f is a contraction mapping.

Finally, from the proof of the above theorem, one observes that the fixed point is obtained by the method of successive approximations (A-1) where the starting value x_0 is completely arbitrary.

BIBLIOGRAPHY

Anasov, D. V.

[1] On Limit Cycles of a System of Differential Equations with a Small Parameter in the Highest Derivative (Russian), *Mat. Sbornik N.S.* **50(92)**(1960), 299–334.

Bailey, H. R., and R. A. Gambill

[1] On Stability of Periodic Solutions of Weakly Nonlinear Differential Systems, *J. Math. Mech.* **6**(1957), 655–668.

Bass, R. W.

[1] Equivalent Linearization, Nonlinear Circuit Analysis and the Stabilization and Optimization of Control Systems, *Proc. Symposium Nonlinear Circuit Analysis*, vol. VI, 1956.

[2] Mathematical Legitimacy of Equivalent Linearization by Describing Functions, IFAC Congress, 1959.

Besicovitch, A. S.

[1] "Almost Periodic Functions," Dover Publications, Inc., New York, 1954.

Blehman, I. I.

[1] On the Stability of Periodic Solutions of Quasilinear Systems with Many Degrees of Freedom (Russian), *Doklady Akad. Nauk S.S.S.R.* **104**(1955), 809–812.

[2] *Ibid.* **112**(1957), 183–186.

Bogoliubov, N.

[1] On Some Statistical Methods in Mathematical Physics (Russian), *Akad. Nauk Ukr. R.S.R.* 1945.

Bogoliubov, N., and Y. A. Mitropolski

[1] Asymptotic Methods in the Theory of Nonlinear Oscillations, Moscow, 1955; revised, 1958 (Russian), *Goz. Iz. Fiz. Mat. Lit.;* translated by Gordon and Breach, 1962.

[2] The Method of Integral Manifolds in Nonlinear Mechanics, Symposium on Nonlinear Vibrations, Kiev, U.S.S.R., September, 1961.

Bogoliubov, N., Jr., and B. I. Sadovnikov

[1] On Periodic Solutions of Differential Equations of the nth-order with a Small Parameter, Symposium on Nonlinear Vibrations, Kiev, U.S.S.R., September, 1961.

[2] Periodic Solutions of a Differential Equation of the *n*th Order with a Small Parameter, *Ukr. Mat. Zh.* **13**(1961), 3, 3–11.

Bohr, H.

[1] "Almost Periodic Functions," Chelsea Publishing Company, New York, 1947.

Boyer, R. C.

[1] Sinusoidal Signal Stabilization, Master's Thesis, Purdue University, January, 1960.

Cartwright, M. L.

[1] Forced Oscillations in Nonlinear Systems, "Contributions to the Theory of Nonlinear Oscillations," vol. 1, pp. 149–241, Annals of Mathematics Studies, No. 20, Princeton University Press, Princeton, N.J., 1950.

Cesari, L.

[1] Asymptotic Behavior and Stability Problems in Ordinary Differential Equations, Ergebn., Heft 16, Springer-Verlag OHG, Berlin, 1959.

[2] Existence Theorems for Periodic Solutions of Nonlinear Lipschitzian Differential Equations, "Contributions to the Theory of Nonlinear Oscillations," vol. 5, Annals of Mathematics Studies, Princeton University Press, Princeton, N.J., 1960.

[3] Functional Analysis and Periodic Solutions of Nonlinear Differential Equations, "Contributions to Differential Equations," vol. 1, Interscience Publishers, Inc., New York, 1962.

[4] Sulla stabilita delle soluzioni dei sistemi di equazioni differenziali lineari a coefficienti periodici, *Atti Accad. Italia, Mem. Classe Fis. Mat. e Nat.* (6)**11** (1940), 633–692.

Cesari, L., and J. K. Hale

[1] Second Order Linear Differential Equations with Periodic L-integrable Coefficients, *Riv. Mat. Univ. Parma.* **5**(1954), 55–61.

[2] A New Sufficient Condition for Periodic Solutions of Weakly Nonlinear Differential Systems, *Proc. Am. Math. Soc.* **8**(1957), 757–764.

Coddington, E. A., and N. Levinson

[1] "Theory of Ordinary Differential Equations," McGraw-Hill Book Company, Inc., New York, 1955.

Demidovich, B. P.

[1] Forced Oscillations of a Quasilinear System in the Presence of a Rapidly Changing External Force, *Prikl. Mat. Mech.* **25**(1961), 705–715; TPMM 1044–1059.

Diliberto, S. P.

[1] Perturbation Theorems for Periodic Surfaces, *Circ. Mat. Palermo* (2)**9**(1960), 265–299

[2] *Ibid.* (2)**10**(1961), 111-112.

Diliberto, S. P., and G. Hufford

[1] Perturbation Theorems of Nonlinear Differential Equations, "Contributions to the Theory of Nonlinear Oscillations," vol. 3, pp. 257–261, Annals of Mathematics Studies, No. 36, Princeton University Press, Princeton, N.J., 1956.

Farnell, A. B., C. E. Langenhop, and N. Levinson

[1] Forced Periodic Solutions of a Stable Nonlinear System of Differential Equations, *J. Math. Phys.* **29**(1951), 300–302.

Favard, J.

[1] "Leçons sur les fonctions presque-périodiques," Gauthier-Villars, Paris, 1933.

Flatto, L., and N. Levinson

[1] Periodic Solutions of Singularly Perturbed Equations, *J. Math. Mech.* **4**(1955), 943–950.

Friedrichs, K.

[1] Special Topics in Analysis, New York University Lecture Notes, 1953–1954.

Fuller, W.

[1] Existence of Periodic Solutions of Weakly Nonlinear Differential and Differential-difference Equations, Ph.D. Thesis, Purdue University, 1957.

Gambill, R. A.

[1] Criteria for Parametric Instability for Linear Differential Systems with Periodic Coefficients, *Riv. Mat. Univ. Parma.* **6**(1955), 37–43.

Gambill, R. A., and J. K. Hale

[1] Subharmonic and Ultraharmonic Solutions for Weakly Nonlinear Differential Systems, *J. Rat. Mech. Ana.* **5**(1956), 353–394.

Gel'fand, J. M., and V. B. Lidskii

[1] On the Structure of the Regions of Stability of Linear Canonical Systems of Differential Equations with Periodic Coefficients, *Uspekhi Mat. Nauk (N.S.)*, **10**(1955), 3–40; *Am. Math. Soc. Transl.* (2)**8**(1958), 143–182.

Glatenok, I. V.

[1] On the Foundation of the Method of Harmonic Balance. Symposium on Nonlinear Vibrations, Kiev, U.S.S.R., September, 1961.

Golomb, M.

[1] Expansions and Boundedness Theorems for Solutions of Linear Differential Systems with Periodic or Almost Periodic Coefficients, *Arch. Rat. Mech. Ana.* **2**(1958), 284–308.

[2] Solutions of Certain Nonautonomous Systems by Series of Exponential Functions, *Illinois J. Math.* **3**(1959), 45–65.

[3] On the Reducibility of Certain Linear Differential Systems, *J. Reine Angew. Math.* **205**(1960–1961), 171–185.

Hale, J. K.

[1] Integral Manifolds of Perturbed Differential Systems, *Ann. Math.* **73**(1961), 496–531.

[2] On Differential Equations Containing a Small Parameter, "Contributions to Differential Equations," vol. 1, Interscience Publishers, Inc., New York, 1962.

[3] On the Behavior of the Solutions of Linear Periodic Differential Systems Near Resonance Points, "Contributions to the Theory of Nonlinear Oscillations," vol. 5, pp. 55–89, Annals of Mathematics Studies, Princeton University Press, Princeton, N.J., 1960.

[4] On the Stability of Periodic Solutions of Weakly Nonlinear Periodic and Autonomous Differential Systems, *Ibid.*, 91–113.

[5] Linear Systems of First and Second Order Differential Equations with Periodic Coefficients, *Illinois J. Math.* **2**(1958), 586–591.

[6] Sufficient Conditions for the Existence of Periodic Solutions of Systems of Weakly Nonlinear First and Second Order Differential Equations, *J. Math. Mech.* **2**(1958), 163–172.

[7] Periodic Solutions of Nonlinear Systems of Differential Equations, *Riv. Mat. Univ. Parma.* **5**(1954), 281–311.

Hale, J. K., and G. Seifert

[1] Bounded and Almost Periodic Solutions of Singularly Perturbed Equations, *J. Math. Ana. and Appl.* **3**(1961), 18–24.

Haupt, O.

[1] Über lineare homogene Differentialgleichungen zweiter Ordnung mit periodischen Koeffizienten, *Math. Annalen.* **79**(1919), 278.

Imaz, C.

[1] On Linear Differential Equations with a Small Parameter, Ph.D. Thesis, University of Mexico, 1961.

Kaplan, K. R.

[1] Design of Intentionally Nonlinear Systems, chap. VI, "Adaptive Control Systems," edited by Mishkin and Braun, McGraw-Hill Book Company, Inc., New York, 1961.

Klotter, K.

[1] Steady State Vibrations in Systems Having Arbitrary Restoring and Arbitrary Damping Forces, *Proc. Symposium Nonlinear Circuit Analysis*, vol. II, 1953, 234–257.

Knobloch, H. W.

[1] Remarks on a paper of L. Cesari on Functional Analysis and Nonlinear Differential Equations. To appear.

Krylov, N., and N. Bogoliubov

[1] "Introduction to Nonlinear Mechanics," Annals of Mathematics Studies, No. 11, Princeton University Press, Princeton, N.J., 1947.

[2] The Application of Methods of Nonlinear Mechanics to the Theory of Stationary Oscillations (Russian), Publication 8 of the Ukrainian Academy of Science, Kiev, 1934.

Kushul, M. Ya.

[1] On Quasiharmonic Systems Which Are Adjacent to Systems with Constant Coefficients in Which Purely Imaginary Roots of the Characteristic Equation Have Nonsimple Elementary Divisors, *Prikl. Mat. Mech.* **22**(1958), 519–533.

Kyner, W. T.

[1] Small Periodic Perturbations of an Autonomous System of Vector Equations, "Contributions to the Theory of Nonlinear Oscillations," vol. 4, pp. 111–125, Annals of Mathematics Studies, No. 41, Princeton University Press, Princeton, N.J., 1958.

Langenhop, C. E.

[1] Note on Almost Periodic Solutions of Nonlinear Differential Equations, *J. Math. Phys.* **38**(1959), 126–129.

Lefschetz, S.

[1] Differential Equations—Geometric Theory, Interscience Publishers, Inc., New York, 1957.

Levinson, N.

[1] Small Periodic Perturbations of an Autonomous System with a Stable Orbit, *Ann. Math.* **52**(1950), 727–738.

Lewis, D. C.

[1] On the Role of First Integrals in the Perturbation of Periodic Solutions, *Ann. Math.* **63**(1956), 535–548.

[2] On the Perturbation of a Periodic Solution When the Variational Equation Has Nontrivial Periodic Solutions, *J. Rat. Mech. Ana.* **4**(1955), 795–815.

Liapunov, A.

[1] Problème générale de la stabilité du mouvement, *Ann. Math. Studies*, No. 17, Princeton, 1947.

Loud, W. S.

[1] Periodic Solutions of $x'' + cx' + g(x) = \epsilon f(t)$, *Mem. Am. Math. Soc.*, No. 31, 1959, 58 pp.

[2] Periodic Solutions of a Perturbed Autonomous System, *Ann. Math.* **70**(1959), 490–529.

[3] Locking-in in Perturbed Autonomous Systems, Symposium on Nonlinear Vibrations, Kiev, U.S.S.R., September, 1961.

[4] The Location of an Invariant Manifold for a Perturbed Autonomous System, *J. Math. Phys.* **40**(1961), 87–102.

Ludeke, C. A.

[1] The Generation and Extinction of Subharmonics, *Proc. Symposium Nonlinear Circuit Analysis*, vol. II, 1953, 215–233.

McHarg, E.

[1] A Differential Equation, *J. London Math. Soc.* **32**(1947), 83–85.

Magnus, W.

[1] The Discriminant of Hill's Equation, NYU, Inst. Math. Sci., Div. Electromagnetic Research, *Research Rept.* BR-28.

Malkin, I. G.

[1] Theory of Stability of Motion, Moscow, 1952, Atomic Energy Commission Translation, AEC-tr-3352.

Mandelstam, L., and N. Papalexi

[1] Über Resonanzerscheinungen bei Frequenzteilung, *Z. Physik.* **73**(1932), 223.

Minorsky, N.

[1] "Introduction to Nonlinear Mechanics," J. W. Edwards, Publisher, Incorporated, Ann Arbor, Mich., 1947.

Mitropolski, Y. A.

[1] On the Investigation of an Integral Manifold for a System of Nonlinear Equations with Variable Coefficients (Russian), *Ukr. Mat. Zhur.* **10**(1958), 270–279.

[2] On the Stability of a One Parameter Set of Solutions of a System of Equations with Variable Coefficients (Russian), *Ukr. Mat. Zhur.* **10**(1958), 389–393.

Moser, J.

[1] New Aspects in the Theory of Stability of Hamiltonian Systems, *Comm. Pure Appl. Math.* **9**(1958), 81–114.

Nohel, J. A.

[1] Stability of Perturbed Periodic Motions, *J. Reine Angew. Mat.* **203**(1960), 64–79.

Oldenburger, R., and R. C. Boyer

[1] Effects of Extra Sinusoidal Inputs to Nonlinear Systems, Paper 61 – Wa—66 presented at Winter Annual Meeting of American Society of Mechanical Engineers, New York, Nov. 26–Dec. 1, 1961.

Opial, Z.

[1] Sur un théorème de A. Filippoff, *Ann. Pol. Math.* **5**(1958), 67–75.

Perlis, S.

[1] "Theory of Matrices," Addison-Wesley Publishing Company, Inc., Reading, Mass., 1952.

Plaat, O.

[1] Synchronization of Coupled Oscillators, Ph.D. Dissertation, University of California, Berkeley, 1958.

Reuter, G. E. H.

[1] Subharmonics in a Nonlinear System with Unsymmetrical Restoring Force, *Quart. J. Mech. Appl. Math.* **2**(1949), 198–207.

Sethna, P. R.

[1] Coupling in Certain Classes of Nonlinear Systems, *Proc. Intern. Symposium Nonlinear Oscillations*, Colorado Springs, 1961.

Sibuya, Y.

[1] Nonlinear Ordinary Differential Equations with Periodic Coefficients, *Funcialaj Ekvacioj* **1**(1958), 121–204.

[2] On Perturbation of Periodic Solutions, *J. Math. Mech.* **5**(1960), 771–782.

Stoker, J. J.

[1] "Nonlinear Vibrations," Interscience Publishers, Inc., New York, 1950.

Urabe, M.

[1] Geometric Study of Nonlinear Autonomous Systems, *Funcialaj Ekvacioj* **1**(1958), 1–83.

Volk, I. M.

[1] On Periodic Solutions of Quasilinear Differential Equations in Certain Cases (Russian), *Izvest. Vyss. Ucebn, Zaved. Matematika* **3**(4) (1958), 31–40.

Volosov, V. M.

[1] On Solutions of Certain Perturbed Systems in the Neighborhood of a Periodic Motion (Russian), *Doklady Akad. Nauk S.S.S.R.* **123**(1958), 588–590.

Wasow, W.

[1] Solutions of Certain Nonlinear Differential Equations by Series of Exponential Functions, *Illinois J. Math.* **2**(1958), 254–260.

Winkler, S., and W. Magnus

[1] The Coexistence Problem for Hill's Equation, NYU, Inst. Math. Sci., Div. Electromagnetic Research, *Research Rept.* BR-26.

Yakubovich, V. A.

[1] Critical Frequencies of Quasi-canonical Systems (Russian), *Vestnik Leningrad Univ.* **13**(1958), 35–63.

[2] On the Dynamic Stability of Elastic Systems (Russian), *Doklady Akad. Nauk S.S.S.R.* **121**(1958), 602–605.

[3] The Small Parameter Method for Canonical Systems with Periodic Coefficients, *Prikl. Mat. Mech.* **23**(1959), 15–34.

[4] Structure of the Group of Symplectic Matrices and the Unstable Canonical Systems with Periodic Coefficients (Russian), *Mat. Sbornik N.S.* **44(86)** (1958), 313–352.

Zadiraka, K. V.

[1] On Periodic Solutions of a System of Nonlinear Differential Equations with a Small Parameter in the Derivatives (Ukrainian), *Dopovidi Akad. Nauk Ukr. R.S.R.*, No. 2, 1958.

INDEX